Geography Education in the Digital World

Geography Education in the Digital World draws on theory and practice to provide a critical exploration of the role and practice of geography education within the digital world. It considers how living within a digital world influences teacher identity and professionalism and is changing young people's lives. The book moves beyond the applied perspective of educational technology to engage with wider social and ethical issues of technology implementation and use of digital data within geography education.

Situated at the intersection between research and practice, chapters draw on a wide range of theory to consider the role, adoption and potential challenges of a range of digital technologies in furthering geographical education for future generations. Bringing together academics from the fields of geography, geography education and teacher education, the book engages with four key themes within the digital world:

- Professional practice and personal identities.
- Geographical sources and connections.
- Geospatial technologies.
- Geographical fieldwork.

This is a crucial read for geographers, geography educators and geography teacher educators, as well as those engaging with existing and new technologies to support geographical learning in the dynamic context of the digital world. It will also be of interest to any students, academics and policymakers wanting to better understand the impact of digital media on education.

Nicola Walshe is Head of the School of Education and Social Care at Anglia Ruskin University.

Grace Healy is the Curriculum Director at David Ross Education Trust.

"In this book Nicola Walshe and Grace Healy have successfully brought together research that critically examines the future of geography education in the digital, and post-digital, worlds. The chapters provide support and guidance for geographers, geography educators, researchers and teacher educators in their efforts to navigate the complexities of the digital world. Particular reference is made to teaching, learning and professional development across all phases of geography education."

– **Graham Butt**, *Emeritus Professor of Education, School of Education, Oxford Brookes University*

"*Geography Education in the Digital World* is timely, thoughtful and wide-ranging. Optimistic in stance, it extends the geospatial into ways in which digital environments influence, reconstruct and normalise how young people and educators do and can construct and use knowledge and act in the world, and especially, why and to what end through schools and higher education. It does not shy away from current and likely challenges but engages with, examines and suggests positive directions from children's everyday contexts to specialist needs in geography teaching and learning. It is a stimulating, forward-looking and provocative book encouraging curriculum thinking and indicating where our (post) digital age is taking geography education."

– **Simon Catling**, *Emeritus Professor of Primary Education, School of Education, Oxford Brookes University*

"*Geography Education in the Digital World* is a must-read for geography educators at all levels who are navigating the digital world as part of their teaching. This book takes readers through the wider issues related to the role of technology in geography education recognising that engaging with digital data requires an understanding of the social relations, cultures, politics and economics of education. This is an important text which moves beyond the applied technicalities of teaching geography through digital data and considers the broader social science concerns of engaging with these digital worlds in geography education contexts."

– **Ruth Healey**, *Associate Professor in Pedagogy in Higher Education, Department of Geography and International Development, University of Chester*

Geography Education in the Digital World

Linking Theory and Practice

Edited by
Nicola Walshe and Grace Healy

LONDON AND NEW YORK

First published 2021
by Routledge
2 Park Square, Milton Park, Abingdon, Oxon OX14 4RN

and by Routledge
52 Vanderbilt Avenue, New York, NY 10017

Routledge is an imprint of the Taylor & Francis Group, an informa business

© 2021 selection and editorial matter, Nicola Walshe and Grace Healy; individual chapters, the contributors

The right of Nicola Walshe and Grace Healy to be identified as the authors of the editorial material, and of the authors for their individual chapters, has been asserted in accordance with sections 77 and 78 of the Copyright, Designs and Patents Act 1988.

All rights reserved. No part of this book may be reprinted or reproduced or utilised in any form or by any electronic, mechanical, or other means, now known or hereafter invented, including photocopying and recording, or in any information storage or retrieval system, without permission in writing from the publishers.

Trademark notice: Product or corporate names may be trademarks or registered trademarks, and are used only for identification and explanation without intent to infringe.

British Library Cataloguing in Publication Data
A catalogue record for this book is available from the British Library

Library of Congress Cataloging-in-Publication Data
Names: Walshe, Nicola, editor. | Healy, Grace, editor.
Title: Geography education in the digital world / edited by Nicola Walshe and Grace Healy.
Identifiers: LCCN 2020018285 | ISBN 9780367224462 (hardback) | ISBN 9780367224479 (paperback) | ISBN 9780429274909 (ebook)
Subjects: LCSH: Geography–Study and teaching. | Educational technology. | Education–Effect of technological innovations on. | Geography–Study and teaching–Audio-visual aids
Classification: LCC G74 .G394 2020 | DDC 910.71–dc23
LC record available at https://lccn.loc.gov/2020018285

ISBN: 978-0-367-22446-2 (hbk)
ISBN: 978-0-367-22447-9 (pbk)
ISBN: 978-0-429-27490-9 (ebk)

Typeset in Galliard
by Taylor & Francis Books

For Ciara, Emma and Rory.

Contents

List of illustrations	ix
Foreword by David Lambert	xii
Acknowledgements	xv
List of contributors	xvi

1 Introduction: Navigating the digital world as geographers and geography educators   1
NICOLA WALSHE AND GRACE HEALY

PART I
Professional practice and personal identities in the digital world   5

2 Teacher identity, professional practice and online social spaces   7
CLARE BROOKS

3 Digital technologies and their roles in knowledge recontextualisation and curriculum making   17
STEVE PUTTICK

4 Navigating the theory-practice divide: Developing trainee teacher pedagogical content knowledge through 360-degree immersive experiences   26
NICOLA WALSHE, PAUL DRIVER AND MANDY-JANE KEENOY

5 Children, childhood and children's geographies: Evolving through technology   38
LAUREN HAMMOND

PART II
Geographical sources and connections in the digital world   51

6 Geographical sources in the digital world: Disinformation, representation and reliability   53
MARGARET ROBERTS

7 'Connecting the Classroom': Teaching geographies of development via digital interactive spaces 65
RORY PADFIELD

8 Social media as a tool for geographers and geography educators 75
FRANCESCA FEARNLEY

PART III
Geospatial technologies in the digital world 87

9 Insights from professional discourse on GIS: A case for recognising geography teachers' repertoire of experience 89
GRACE HEALY

10 Empowering geography teachers and students with geographical knowledge: Epistemic access through GIS 102
MARY FARGHER AND GRACE HEALY

11 GIS for young people's participatory geography 117
SUSAN PIKE

PART IV
Geographical fieldwork in the digital world 129

12 Using mobile virtual reality to enhance fieldwork experiences in school geography 131
REBECCA KITCHEN

13 Teaching and learning geography with mobile technologies and fieldwork 142
CHEW-HUNG CHANG

14 Augmented reality: Opportunities and challenges 155
GARY PRIESTNALL

15 Location-based games for geography and environmental education 168
STEFFEN SCHAAL

PART V
Conclusion 179

16 From the digital world to the post-digital world: The future generation of geographers 181
GRACE HEALY AND NICOLA WALSHE

Index 186

Illustrations

Figures

2.1	Key components of a socio-cultural ecological approach to mobile learning. From Pachler et al. (2010, p.4). Reprinted by permission.	12
5.1	Harvey's grid of spatial practices.	43
6.1	Illustration of how a lesson activity can include all parts of the participant dimension from the 'Storm Chasers' lesson.	59
7.1	Learning outcomes of the 'Connecting the Classroom' exercise.	70
10.1	Mapping relationships between earthquake magnitude and depth in ArcGIS Online.	106
10.2	Identifying sites in King's Cross for environmental quality fieldwork data with Survey 123.	108
13.1	Word cloud showing occurrence of words about mobile technology and fieldwork within abstracts from articles in Annex A (articles from *International Research in Geographical and Environmental Education* and *Journal of Geography*).	144
13.2	The TPACK framework.	146
14.1	Augmented scenes using locative media.	161
14.2	Projection Augmented Relief Model (PARM) supporting in-class discussion of flood risk.	163
15.1	Areas for geogame design (adapted from Schaal and Schaal, 2018, p.44).	172
15.2	Sequencing of location-based tasks.	173

Tables

6.1	The participation dimension.	58
6.2	Extract from the EU's DigComp 2.0: The Digital Competency framework for Citizens. Component 1: Information and Data Literacy.	61
7.1	Project topic, background and key research questions for the 'Connecting the Classroom' exercise.	69
9.1	Journal articles from *Primary Geography* (PG) and *Teaching Geography* (TG) with a focus on GIS.	93
10.1	Maude's (2016) typology of powerful geography knowledge adapted by Lambert and Solem (2017, p.11).	105

10.2	Powerful geographical knowledge developed through analysing earthquake patterns with ArcGIS Online (Example 1).	107
10.3	Powerful geographical knowledge developed through collecting and analysing environmental quality data with Esri Survey 123 (Example 2).	109
10.4	Powerful geographical knowledge developed through investigating fracking in QGIS Cloud (Example 3).	110
10.5	Powerful geographical knowledge developed using an Esri StoryMap to study Danakil, Ethiopia (Example 4).	111
13.1	Frequency of occurrence of words within article titles about mobile technology and fieldwork in *International Research in Geographical and Environmental Education* and *Journal of Geography*.	143
14.1	Categories of augmented reality.	156

Foreword

The influential *New York Times* columnist Thomas Friedman, the man who brought us *The World is Flat* in 2007 (thankfully, not in support of the Flat Earth Society), has argued passionately about the 'age of accelerations' (Friedman, 2016). Take the climate emergency for example, caused by the exponential growth in the deposition of carbon dioxide into the atmosphere since the industrial revolution and especially in the last quarter of the 20th century. The climate is an emergency *because of* the accelerating impacts of fossil fuel consumption: rising global temperatures, extreme weather events, the melting of Greenland ice, rising sea levels, and release of methane from beneath the thawing permafrost (etc.).

Another of Friedman's accelerations is the development of technologies, including digital technologies. Friedman shows a graph to demonstrate that the rate of technological change, which was slow for most of human history, has also now accelerated to such a dizzying speed that it now occurs at a rate (he argues) that outpaces human ability to adapt to it. This could lead us to a dystopian world view of human beings controlled by and at the mercy of machines that they do not understand. But Friedman is not a pessimist. He explains how it falls to education (and other institutions in society) to respond. We can make up the gap by preparing young people in schools more effectively to adapt, to be more resilient, to accept flexibility and to embrace change. This of course brings us to the familiar discourse of 'lifelong learning', '21st century skills' 'transversal competences' and 'learning to learn', aspects of what Gert Biesta (2015) has labelled (disparagingly) the neoliberal 'learnification' of education. And despite Friedman's infectious optimism, those interested in a quality of education which is truly fit for the century and the various challenges of the human epoch must accept that finding an appropriate response is far from straightforward.

So, when it comes to the consideration of geography in education and how this relates to and/or utilises the affordances of the digital world, we know we are in for some fun! As the title of this book implies, the 'digital world' envelops all parts of geography education. Even if teachers or pupils wanted to ignore or escape digital technologies it would not be possible. This reminds me of something I think is attributable to Seymour Papert: that you know a technology is working when you are no longer really conscious of it. Like coloured pencils. Great invention, and no-one now thinks twice about them. We can bring this up to date – with email for example, or the internet: how quickly these technologies have become (to all intents and purposes) ubiquitous. It is a legitimate question to ask *how* these technologies are working. For example, how has the advent of Twitter affected, say, the quality of political discourse in democratic societies? The refusal of leaders to accept challenging interviews from experienced and knowledgeable journalists and substitute this with regular social media utterances – which then become 'the news' – is almost certainly undermining democratic principles of openness and transparency. And this of course is ironic, for the promise of the internet has always been openness and transparency, based upon the rationale that 'we' now all have a voice.

Society-wide therefore, technological accelerations present self-evidently serious issues and questions. Within the specialist field of geography education, we cannot take on these matters entirely ourselves. But it is important to make a considered contribution to wider educational debates. It is not a matter of embracing technologies or resisting them. We are way beyond those kinds of considerations. They are here in our lives and educationists need to have a view on what impact they may have on education – indeed, how certain technologies may themselves shape how we think of education itself. Geography in education has a reputation for being an early adopter of technologies – after all, it is an information-rich subject with an intrinsic interest in a rapidly changing world, and information is now instantly available at the touch of a button; indeed, we are soaked in information. It is also multi-modal in its research and communication technique, using images of all kinds often derived from the burgeoning GI industry. It concerns human behaviour and for decades has therefore explored gaming strategies to understand decision-making within the context of economic, political, environmental and social processes. Geography educators therefore need to deeply consider the potentials – and the risks – of digital technologies for teaching and learning both within and beyond the geography classroom.

It is therefore brilliant to see this book, and I have enjoyed reading it. The project was conceived during a regular GEReCo[1] discussion of research priorities in geography education during my stint as Chair (2015–2019), and so I have been expecting its arrival – and my expectations have been exceeded. Nicola Walshe and Grace Healy have brought together a wide-ranging and unique collection of papers addressing all stages of education, including teacher education. The book as a whole is not only explanatory and laced with case studies and examples but also suggestive of where research effort may usefully be directed now and in the future.

One thread running through the book raises questions concerning the changing *role* of the teacher – and some authors are understandably (and correctly) far from decisive about this. The book is successful in opening up this discussion, contributing to our capacity to respond to what Alvin Toffler, some years ago, called *Future Shock*:

> To survive, to avert what we have termed future shock, the individual must become infinitely more adaptable and capable than ever before. We must search out totally new ways to anchor ourselves, for the old roots – religion, nation, community, family or profession – are now shaking under the hurricane impact of the accelerative thrust. (Tofler, 1971, p.35)

Teachers – and students – need to be able to maintain agency in this scenario. Walshe and Healy's book contributes and provides a step forward in our collective endeavour to work out how to do this.

David Lambert
Emeritus Professor of Geography Education at the UCL Institute of Education, London, UK

Note

1 This is the UK-based Geography Education Research Collective (www.gereco.org). In 2019 this group merged with the British sub-committee of the International Geographical Union Commission on Geographical Education.

References

Biesta, G., 2015. What is education for? On good education, teacher judgement and educational professionalism. *European Journal of Education*, 50 (1), pp.75–87.
Friedman, T., 2016. *Thank You for Being Late: An Optimist's Guide to Thriving in the Age of Accelerations*. New York: Macmillan.
Tofler, A., 1971. *Future Shock*. New York: Random House (Bantam edition).

Acknowledgements

The development of this book was supported by GEReCo, the Geography Education Research Collective; the editors would like to thank the GEReCo committee for their support and guidance. Particular thanks go to Professor David Lambert without whose encouragement this book would not have come to fruition, and Dr Clare Brooks for invaluable advice during the planning and editing stages.

The editors share their sincere gratitude with the chapter authors who contributed their time and expertise to this book.

We are also very grateful to the following individuals, groups and organisations for granting permission to reproduce material in the following chapters:

Chapter 2. IGI Global, for the diagram from: Pachler, N., Cook, J. and Bachmair, B., 2010. Appropriation of mobile cultural resources for learning. *International Journal of Mobile and Blended Learning*, 2, pp.1–21.

Chapter 5. Yale School of Architecture, for Harvey's grid of spatial practices from: Harvey, D., 1990. Flexible accumulation through urbanisation reflections on 'post-modernism' in the American city. *Perspecta*, 26, pp.251–272.

Chapter 6. Ellena Mart, International School of Geneva, for the illustration of how a lesson activity can include all parts of the participant dimension from: Storm Chasers (extracts from website Geogalot: www.geogalot.com).

Chapter 6. Joint Research Centre, the European Commission's in-house science service, for the extract from the EU's DigComp 2.0: The Digital Competency Framework for Citizens (Component 1: Information and Data Literacy) from: Vuorikari, R., Punie, Y., Carretero Gomez, S. and Van den Brande, G., 2016. *DigComp 2.0: The Digital Competence Framework for Citizens. Update Phase 1: The Conceptual Reference Model* (EUR 27948 EN). Luxembourg: European Union. Available at: https://publications.jrc.ec.europa.eu/repository/bitstream/JRC101254/jrc101254_digcomp%202.0%20the%20digital%20competence%20framework%20for%20citizens.%20update%20phase%201.pdf [Accessed 7 January 2020].

Chapter 10. AGTA, for the table from: Lambert, D. and Solem, M., 2017. Rediscovering the teaching of geography with the focus on quality. *Geographical Education*, 30, pp.8–15.

Chapter 10. Esri maps created using ArcGIS® software. ArcGIS® and ArcMap™ are the intellectual property of Esri and are used herein under license. © Esri. All rights reserved. For more information about Esri® software, please visit www.esri.com

Chapter 13. Matthew Koehler and Punya Mishra, of Michigan State University for permission to use the TPACK image from the original source http://tpack.org

Chapter 14. The Geography Department at Keswick School, Cumbria, for the photograph of students using a projection augmented relief model (PARM) to explore flood risk.

Contributors

Clare Brooks is a Reader in Geography Education at the UCL Institute of Education (IOE). Until recently, she was Head of the Academic Department of Curriculum, Pedagogy and Assessment, and Head of Initial Teacher Education. Prior to undertaking these roles, Clare led a range of Geography Education programmes at the IOE. She is a founding member and Chair of GEReCo (Geography Education Research Collective) and is Co-Chair of the International Geographical Union – Commission for Geography Education (IGU-CGE). Clare's research interests are in teacher education and development, particularly around teacher subject identity. Her latest book is *Teacher Subject Identity in Professional Practice* published by Routledge, and she is Co-Editor of the Commission's book series on International Perspectives on Geography Education published by Springer.

Chew-Hung Chang is the Chief Planning Officer at the National Institute of Education, Nanyang Technological University, Singapore, looking after academic quality, innovations in teaching and strategic planning for the Institute. He is also concurrently an Adjunct Professor of East China Normal University, Shanghai, China. He has published widely in the areas of geography, geographical education and environmental education, with a particular focus on climate change education and the geography curriculum in schools. He is currently the Co-Chair of the International Geographical Union – Commission on Geography Education (IGU-CGE), co-editor of the journal *International Research in Geographical and Environmental Education*, as well as the President of the Southeast Asian Geography Association. Chew-Hung is also the International Editor of the journal *Review of International Geographical Education online (RIGEO)* for the Asia and Oceania region, as well as serving as a member of the scientific committee of the journal *J-Reading – Journal of Research and Didactics in Geography*.

Paul Driver is a Senior Learning Technologist at Anglia Ruskin University, specialising in immersive educational virtual reality scenario creation. He holds an MA in Creative Media Practice in Education, has over 25 years of teaching experience and is an award-winning international educational materials writer, a teacher trainer and book illustrator. Paul has been nominated for four Teaching Innovation Awards and, in 2016, received the English-Speaking Union Award for his teacher resource book, *Language Learning with Digital Video*. In 2018, he received the National Learning Technologist of the Year award. Paul's research interests span across many fields, exploring the roles of technology, pedagogy, game design, play, and embodied cognition in the process of learning.

Mary Fargher is Programme Leader for the MA in Education (Geography) at UCL Institute of Education, London. Prior to her current roles, Mary was a secondary school geography teacher and leader in schools and colleges in London. Her main research interest is the use of GIS to support high-quality geography education. She has recently completed a collaborative research project with maths and science colleagues on the role of master's students working together to support each other in their research.

Francesca Fearnley is a recent graduate from the School of Geography at Plymouth University and current postgraduate at the University of Gloucestershire. It was during the process of writing her undergraduate dissertation 'Tracking ISIS attacks through Twitter' that her interest grew in the use of social media, and Twitter in particular, as a source of information for geographical study. Since graduating, Francesca has continued to explore the subject with further articles and publications, including in the Summer 2018 edition of *Geography*.

Lauren Hammond is Lecturer in Geography Education at the UCL Institute of Education (IOE). A former secondary school teacher in the UK and Singapore, Lauren worked as Subject Lead in a London school prior to joining the IOE. Her primary research interests are in Children's Geographies, with her doctorate examining how, and why, they are of value to pre-university geography education. In addition to this, Lauren also has a keen interest in the geographical concepts of space and place, and also in the relationship(s) between different spaces of geographical thought. Lauren acts as a tutor on the MA in Geography Education, as well as teaching on a number of Initial Teacher Education (ITE) programmes including the Secondary Geography PGCE and Teach First; she also works to develop mentors, and mentoring, across ITE at the IOE.

Grace Healy is the Curriculum Director at David Ross Education Trust. Prior to undertaking this role, she led the geography subject community across a multi-academy trust of 13 primary and secondary schools, and contributed to the leadership of a SCITT. She is currently undertaking a PhD at UCL Institute of Education. Her PhD research explores the role of disciplinary knowledge within geography teachers' professional practice. Grace chairs the Teacher Education Special Interest Group of the Geographical Association, is a member of the Geography Education Research Collective (GEReCo), and has been elected onto the Council of the Royal Geographical Society as Honorary Secretary (Education). She also serves on British Educational Research Association's (BERA) Publication Committee and on the editorial boards of the *Curriculum Journal* and the *London Review of Education*.

Mandy-Jane Keenoy is a third-year undergraduate student on the BA (Hons) Primary Education Studies course at Anglia Ruskin University. She has participated in a number of research studies across her degree course, in particular working as a student research intern for a four-week project on the use of 360-degree immersive experiences in initial teacher education (ITE); the work for this formed the basis of her contribution to this book. Mandy has a wealth of experience working with children through her role as a teaching assistant in a primary school. She has been offered a place on the Primary PGCE course at the University of Cambridge and aims to go on to become a primary school teacher.

Rebecca Kitchen is Continued Professional Development (CPD), Curriculum and Marketing Manager at the Geographical Association, having previously taught and been Head of

Geography for 16 years at a secondary school in Buckinghamshire. Her research interests are focused on student representations and experiences of geography and she has wide-ranging experience in both developing curriculum materials and designing and leading teacher CPD, both in the UK and internationally. Rebecca was involved in the Open University project described in her chapter and has subsequently developed some supporting materials for Google Expeditions based lessons for Twig World.

Rory Padfield is an environmental social scientist working as a Lecturer in the School of Earth and Environment at the University of Leeds following academic positions at Oxford Brookes University (2016–2019) and Universiti Teknologi Malaysia (2010–2016). In his teaching and research Rory engages in the broad themes of uneven development in the Global South, natural resource governance, business practices and sustainable management of supply chains, and media and digital geographies. Rory draws on a variety of academic perspectives to pursue critical questions around ethics, social and environmental responsibility, and local–global political economy. In recent years, Rory has developed a keen interest in exploring the way in which digital technologies can support the internationalisation of higher education in the UK and international contexts.

Susan Pike coordinates Geography Education in the Institute of Education, Dublin City University, Ireland. She teaches and supervises research in Geography, Environmental and Citizenship Education, as well as Outdoor Learning, at undergraduate and postgraduate levels. Susan holds a degree in Geography from the University of Nottingham and an MA (Education) in Geographical and Environmental Education from the University of Southampton. Her doctoral thesis, at the Queen's University Belfast, was an in-depth investigation into children's experiences and participation in their local environments. Susan was the lead author of *EcoDetectives* (2011), a resource for schools for the Department of the Environment and Local Government, and recently completed *Learning Primary Geography: Ideas and Inspiration from Classrooms* (2016) which draws on many examples of geographical and environmental education in schools. Susan is an active member of the Geographical Association, and was President of the organisation in 2020.

Gary Priestnall is an Associate Professor within the School of Geography at the University of Nottingham. His research interests focus on digital geographic representation in a range of contexts and utilising a range of technologies, including virtual and augmented realities and location-aware mobile devices. Gary developed various teaching and learning settings for digital geographic information as Director of the Nottingham arm of the Spatial Literacy in Teaching (SPLINT) Centre for Excellence in Teaching and Learning. His key research outputs include: 'Projection augmented relief models (PARM): Tangible displays for geographic information' (*Proceedings from Electronic Visualisation and the Arts*, 2012); 'Virtual geographic environments' (*Teaching Geographical Information Science and Technology in Higher Education*, 2012); and 'Spatial frames of reference for literature using geospatial technologies' (*Literary Mapping in the Digital Age*, 2016).

Steve Puttick is Associate Professor of Teacher Education at the University of Oxford, and fellow of St Anne's College. He is a curriculum tutor for the Geography PGCE and MSc in Learning and Teaching. Steve is a qualified geography teacher and was previously the Head of Department at a comprehensive secondary school in Oxfordshire. His doctoral research was funded by an ESRC studentship and explored geography teachers' subject knowledge through ethnographic study of three secondary school geography departments. His research interests include the relationships between academic disciplines and school subjects, and teachers' roles in the recontextualisation of knowledge.

Margaret Roberts taught geography in comprehensive schools in London, Leicestershire and Sheffield and was a member of the Nuffield Foundation Resources for Learning Project before going on to lead the geography PGCE course at the University of Sheffield. Her research has focused on interpretations of the Geography National Curriculum and approaches to learning. Since her retirement in 2006, Margaret has continued to write for publication, present papers at conferences (including in Australia, Singapore, Portugal, Italy and the UK) and contribute to Continued Professional Development (CPD). For example, in 2011 she led ten courses on geographical 'inquiry' in Singapore. Margaret's publications include journal articles, chapters in books and two books: *Learning through Enquiry* (2003) and *Geography through Enquiry: Approaches to Teaching and Learning in the Secondary School* (2013). She has been involved in the work of the Geographical Association for many years: on committees, as Editor of *Teaching Geography* and as President (2008–2009). She was Chair of COBRIG (Council of British Geography) from 2011 to 2019.

Steffen Schaal is Full-Professor at the Institute of Natural Sciences and Technology Education at the University of Education Ludwigsburg (LUE) in Germany. After receiving a degree as a secondary school teacher (subjects: biology, French, physical education) he worked as a teacher for several years before gaining an MA in Educational Sciences, Biology and Physical Education and then a PhD in Educational Research (topic: hypermedia-assisted learning in science). Steffen was appointed as Assistant Professor at LUE in 2006, established the Science Education Department as Full-Professor at the University of Bamberg in 2009 and then was appointed Full-Professor at LUE in 2011. Since then he has become Dean of Study Affairs at the Faculty for Natural and Cultural Sciences and leads several projects dealing with media-assisted learning in Science Education and Education for Sustainable Education. His group at LUE works intensively on location-based learning, simulations and AR/VR in Science Education.

Nicola Walshe is Head of the School of Education and Social Care at Anglia Ruskin University. Previously she gained a PhD in Glaciology and taught and worked as Head of Geography in three secondary schools in the UK before going on to teach and lead the Geography PGCE course at Cambridge University. Nicola is Secretary of GEReCo (Geography Education Research Collective) and co-convenor of the Environmental and Sustainability Education Research (ESER) network in the European Educational Research Association (EERA). Her research interests include environmental and sustainability education, pedagogies at the intersection of nature, the arts and wellbeing, and curriculum and pedagogy within teacher education more broadly.

Chapter 1

Introduction

Navigating the digital world as geographers and geography educators

Nicola Walshe and Grace Healy

Education is shaped evermore by the digital world within which we live, yet technology is increasingly something that 'gets done' in education without much thought or reflection (Selwyn, 2013). The explosion of digital media and readily accessible big data opens up opportunities for teaching, learning and professional development (Reyna et al., 2018), but there is a danger of us engaging with the data with little understanding of its provenance. More often than not, discussions about technologies within education, and geography education in particular, gravitate towards a similar set of issues, in particular institutional or individual barriers to engagement with technology (e.g. Walshe, 2017). This also tends to be based around deterministic assumptions that technologies possess inherent qualities that have predictable and positive effects on learning, if used in the correct manner. However, Selwyn (2017) observes that these barriers are not technical issues, but are rather rooted in the social relations, cultures, politics and economics of education.

With this in mind, this book aims to engage with the big picture of geography education in the digital world; it moves beyond the applied perspective of technology and brings the concerns of social science to bear on the use of technology and digital data in geography education contexts. In response to Brooks' (2016, p.21) question 'What does it mean to be a [geography] teacher in this day and age?' this book explores the role of technology within teachers' professionalism and subject expertise, highlighting the significance of technology in shaping geography in the classroom and engaging with mechanisms through which geography educators might be more proactive in drawing on technologies to support both their own role in knowledge recontextualisation and curriculum-making, and their students' geographical learning. At the beginning of this new decade in which geography seems increasingly important for young people, *Geography Education in the Digital World* explores the wider issues and questions as to the role of technology in geography education and how living within a digital world is influencing teacher identity and professionalism, and changing young people's lives.

Geography Education in the Digital World is situated at the intersection between research and practice, drawing on a wide range of theory to consider the role, adoption and potential challenges of a range of digital technologies in furthering geographical education for future generations. It is significant for two reasons: firstly, it explores the relationship between identity and technology, considering themes such as how access to technology has altered young people's lived geographies, teachers' identities in online spaces, and the challenges of adopting new technology into teacher practice. Secondly, it addresses a gap in the current geography education literature, exploring how a range of 'new' technologies, such as geogames and virtual reality, might be used to support geographical learning in and beyond traditional notions of 'the classroom', and considering both practical and ethical dimensions of such practice. As such, this book is for geographers, geography educators and geography teacher educators across schools (primary and secondary) and higher education (HE) institutions that are engaging with new and existing technologies to support geographical learning, as well as those wanting to better understand its influence on geography education per se. The book brings together voices that represent these different phases of education, offering perspectives from those with a range of professional identities, encompassing geography education researchers, postgraduate geographers, teacher educators, geography teachers and those involved with our geography subject association in the UK. It explores primary, secondary and HE contexts, and draws on the expertise of early career researchers through to retired members of the geography education community. We have deliberately incorporated perspectives from those working at the boundaries of geography, such as environmental and social sciences, natural sciences and technology education, as well as including authors working within a range of geographical settings, such as the UK, Germany, Ireland and Singapore. The result is a wide-ranging exploration of *Geography Education in the Digital World*, drawn from multiple perspectives and situated across a variety of contextual boundaries.

The book comprises fourteen chapters organised into four key parts, each of which explores a separate theme pertinent to *Geography Education in the Digital World*. Within Part I, **Professional practice and personal identities in the digital world**, authors consider the impact of the digital world broadly on geography, education and geography education in four chapters. In the first, Clare Brooks considers how the use of online social spaces, including for teacher-to-teacher interaction and support, might affect geography teachers' subject identity and professional practice. Clare describes teachers' harvesting of resources from a range of online sources as a form of *curation*, arguing that the notion of curation can nourish teacher identity as it enforces the importance of professional values, and that which teachers consider to be important. In the second, Steve Puttick continues to consider teacher engagement with digital technologies, revisiting debates surrounding the challenges of technology and reframing them in the light of teachers' knowledge recontextualisation and curriculum-making. In the third chapter, Nicola Walshe, Paul Driver and Mandy-Jane Keenoy consider how immersive 360-degree video technology might be used to navigate the theory-practice divide, particularly as a mechanism for supporting trainee teachers to understand the intersection between geography knowledge and pedagogical understanding, or pedagogical content knowledge. In the final chapter of this part, Lauren Hammond then critically examines how access to technology (specifically Web 2.0) has altered children's geographies, drawing out implications for how, and why, these changes (and children's geographies more broadly) are of value to geography education in schools.

Part II explores **Geographical sources and connections in the digital world**, where authors grapple with how the digital world is changing what lies within the purview of geography educators and students, and the implications for enhancing geographical education of these future generations of geographers. In the first chapter in this part, Margaret Roberts reflects on geographical sources in the digital world, in particular exploring concerns of disinformation, and consideration of representation and reliability of sources; Margaret critically examines use of various sources within geography education, asking what makes high-quality geographical sources and how they can best support geographical learning in and beyond the classroom. In the second chapter in this part, Rory Padfield draws on work undertaken in university geography contexts to consider virtual learning communities, with a specific focus on their potential for building inter-school and HE partnerships. In particular, Rory reveals that this approach to supporting partnership provides meaningful engagement between students from Indonesia and the UK, allowing them to engage critically with their assumptions about development and, thereby, using technology to support decolonisation of knowledge within the classroom. In the final chapter in this part, Francesca Fearnley considers the use of social media as a tool for geographers and geography educators; drawing on contemporary examples from both human and physical geography, Francesca argues that social media presents a range of valuable opportunities for future generations of geographers to learn new skills, develop their repertoire of methods for geographical research, and ask questions that enhance their capacity to think geographically.

Part III comprises three chapters exploring **Geospatial technologies in the digital world**, in particular the use of geographical information systems (GIS). While there is general consensus within the academy that GIS has a place within school geography (e.g. Bednarz, 2004; Kerski et al., 2013; Bearman et al., 2016), and national curricula across the world reflect this, research suggests that development of GIS within school contexts remains inconsistent (e.g. Walshe, 2017). Within this context, the chapters in this part explore three very different dimensions of GIS. In the first, Grace Healy considers insights from professional discourse on GIS within the UK context as a mechanism for exploring geography teachers' repertoire of experience; Grace argues that teachers and researchers need to work collectively to capitalise upon the opportunities and address the challenges of GIS. In the second, Mary Fargher and Grace Healy analyse the potential of WebGIS in the development of geographical knowledge, ultimately making a case to consider how teachers can embed GIS within their repertoire of practice, so as to ensure students are not denied their curriculum entitlement to GIS or the epistemic access afforded by geospatial technologies. In the final chapter, Susan Pike considers GIS for young people's participatory geography, drawing on work across primary and secondary geography schools in Ireland, and highlighting the potential for GIS to develop students' informed perspectives about local geographical issues and support active citizenship.

Within Part IV, four chapters explore **Geographical fieldwork in the digital world**, in particular considering the extent to which developing technologies support students' geographical learning through fieldwork. In the first chapter, Rebecca Kitchen considers the potential of using mobile virtual reality, particularly Google Expeditions, to enhance fieldwork experiences in school geography. Rebecca explicates how virtual reality can help physical fieldtrips to become embedded within students' geographical education, so that fieldwork becomes much more than a one-off experience. In the second, Chew-Hung Chang looks broadly at mobile technologies and fieldwork, arguing how and why the use of mobile technologies and learning in the field need to be developed in tandem, and illuminating the gaps for future research and practice, including cross-fertilisation of ideas between school and academic geography. In the third chapter in this part, Gary Priestnall draws on experience from university-level geography teaching to consider the opportunities and challenges of augmented reality for school geography, particularly supporting children's understanding of maps and landscapes through techniques such as digital terrain modelling. Within the final chapter of this part, Steffen Schaal presents a detailed account of how the use of location-based games can develop geographical and environmental learning, with a particular focus on field-based technologies.

Within the concluding chapter of the book in Part V, we draw on the notion of teachers' curriculum-making (Lambert and Biddulph, 2015), developed throughout many of the earlier chapters, to explore the way that the digital world shapes student, teacher and school subject within the context of the discipline of geography. This includes how teachers ought to be empowered as professionals to navigate this digital world in their curriculum-making, as well as how they might help students navigate this digital world themselves. We finish by questioning how geography education might need to prepare for the post-digital world in which the digital is inextricably integrated within teachers' and students' everyday lives.

References

Bearman, N., Jones, N., André, I., Cachinho, H.A. and DeMers, M., 2016. The future role of GIS education in creating critical spatial thinkers. *Journal of Geography in Higher Education*, 40 (3), pp.394–408.

Bednarz, S.W., 2004. Geographic information systems: A tool to support geography and environmental education? *GeoJournal*, 60 (2), pp.191–199.

Brooks, C., 2016. *Teacher Subject Identity in Professional Practice: Teaching with a Professional Compass*. London: Routledge.

Kerski, J.J., Demirci, A. and Milson, A.J., 2013. The global landscape of GIS in secondary education. *Journal of Geography*, 112 (6), pp.232–247.

Lambert, D. and Biddulph, M., 2015. The dialogic space offered by curriculum-making in the process of learning to teach, and the creation of a progressive knowledge-led curriculum. *Asia-Pacific Journal of Teacher Education*, 43 (3), pp.210–224.

Reyna, J., Hanham, J. and Meier, P., 2018. The internet explosion, digital media principles and implications to communicate effectively in the digital space. *E-Learning and Digital Media*, 15 (1), pp.35–52.

Selwyn, N., 2013. *Education in a Digital World: Global Perspectives on Technology and Education*. Abingdon: Routledge.

Selwyn, N., 2017. *Education and Technology: Key Issues and Debates*. London: Bloomsbury.

Walshe, N., 2017. Developing trainee teacher practice with geographical information systems (GIS). *Journal of Geography in Higher Education*, 41 (4), pp.608–628.

Part I

Professional practice and personal identities in the digital world

Chapter 2

Teacher identity, professional practice and online social spaces

Clare Brooks

Introduction

Online social spaces are ubiquitous now for many people, not just teachers. Different sectors of society have found ways of using various online platforms for different forms of engagement, communication and interaction. Online spaces have been particularly useful for professional networks, especially for professionals who consider themselves isolated or peripheral. As such, for many new teachers engaging in online social spaces is as natural as hanging out in the staff room, or socialising 'down the pub'. However, we also know from public accounts of trolling and cyber-bullying that people can behave differently online. The public nature of online social spaces can lead to professional 'performance' of attitudes and opinions, as well as practices. Sharing and collaboration are also easier, facilitated through minimum-fuss technology. But to what extent does this change our sense of teacher identity and how can the profession respond? In this chapter, I offer some thoughts on both these questions, and suggest that the notion of curation can be a useful way forward.

In my previous work I have argued that teacher subject identity is such an important part of teachers' professional practice that it can operate like a professional compass, giving teachers a sense of direction which can help them to navigate the professional knowledge landscape of today's education climate. I argued that 'Teacher identity is indeed central to understanding how teachers adapt to reform' (Brooks, 2016, p.4). In the narratives that I drew upon in that work, the reforms that the teachers described were mostly related to changes in their school or education context. The types of reforms described by the teachers in my study will be familiar to many teachers: increased accountability, classroom surveillance, school policies that seek to control and deprofessionalise teachers' work. I argued that in order for teachers to navigate such a landscape of reform they needed to have a strong sense of their purpose and their subject story to guide their professional practice (Brooks, 2016).

More recent work into the professional practice of teachers has noted a much broader conception of reform, and indeed how change in the broader society has influenced teacher work. David Mitchell's (2016) work is an excellent example of this approach. In his account of the work of geography departments across North London, he meticulously details their working practices. In this analysis he notes the significant relationships engendered by the geography team and in particular influenced by the Head of Geography, and how such relationships reflect different ways of working and sharing resources across and between geography teachers. His use of the term 'hyper-socialised' reflects the ways in which collaboration between teachers takes place in a number of contexts and ways, and notes in particular the online nature of some of those connections. The changing nature of how teachers work and interact with each other could have significant implications for how teachers view their work and by extension their identity as teachers. So how does that hyper-socialisation affect teacher identity and work: to what extent do these external factors influence how we 'make teachers up'?

The importance of teacher identity

Teacher identity is a key way of understanding teachers' work. Whilst large-scale surveys like the VITAE project reveal general patterns and trends in teachers' working lives (Day et al., 2007; Sammons et al., 2007; Day and Gu, 2010), how the ebb and flow of that work affects individuals will depend on their identity as teachers: the extent to which they feel committed to teaching, how resilient they are to change and the extent to which they are, somewhat simply put, happy in their work.

A focus on teacher subject identity is an important dimension in understanding how teachers' work can act as both a sustaining and moral perspective throughout their career. Ball and Goodson (1985) highlight that teacher identity becomes invested in particular aspects or facets of the teaching role. For many secondary teachers their subject specialism plays a crucial part. They see themselves as scientists, geographers, historians, mathematicians, etc. (Ball and Goodson, 1985).

Ball and Goodson (1985) explain that this may be due to the socialisation that takes place in their initial teacher education and is akin to subject 'sub-cultures' – a shared set of subject-specific values and norms. However, they also acknowledge that some teachers do not fit this category and see themselves as educators rather than scholars, which they describe as those who are 'vocationally committed':

> They see their interests as primarily with caring for the pupils and encouraging their intellectual development. In contrast professionally committed teachers are much more likely to see themselves as subject specialists and to see their subjects as providing an avenue for advancement as well as a source of personal satisfaction. (Ball and Goodson, 1985, p.21)

Regardless of the identity that teachers adopt for themselves, it would be wrong to suggest that this is purely an individual endeavour; as outlined by Ball and Goodson (1985), it is a result of interactions with their environment and their colleagues. To this end, Clandinin and Connelly's (1995) metaphor of a professional knowledge landscape is a useful way of conceptualising how contexts can influence teachers.

Professional knowledge landscapes

The idea that teachers work within a professional knowledge landscape was first introduced by Clandinin and Connelly (1995). They describe these landscapes as being made up of 'sacred stories' which come from official discourses, and which may be in conflict with teachers' own stories of their professional practice. Clandinin and Connelly (1995) argue that this creates knowledge dilemmas as they represent a conflict between official knowledge as represented by policies or directives from their school or official discourses, and the personal knowledge of the teacher. These landscapes can play a profound role in affecting how teachers work, and can, in effect, 'make teachers up' in that they can influence not just what teachers do but also who they are (see Brooks, 2016).

One of the ways that teachers can respond to such 'sacred stories' is to draw upon dimensions of their own professional identity which reflect their values about education and teaching, many of which can stem from their subject identity or how they identify with the broader community of teachers. In other words, teacher identity can be an important driver in how teachers respond to reforms in education. However, this is not just one-way traffic, as context and identity are mutually constructed: teachers make up the contexts they work in, and in turn their teacher identity can be influenced by their place of work.

The research that underpins Clandinin and Connelly's (1995) work tends toward focusing on localised and individual experiences of education contexts (as is true of my own work). These accounts are useful in seeing the impact of reform on the individual. But sometimes these individualised accounts mask bigger changes happening which can influence education; in particular on teachers' work and wellbeing. Some have attributed this to the pervasiveness of neoliberal education policies on how schools operate (Ball et al., 2012) and on the nature of how individual teachers work (Day et al., 2007). Clandinin and Connelly's (1995) metaphor highlights that professional knowledge landscapes are both moral and knowledge-based landscapes, which can reflect the dominant discourses in education, and so can situate teachers as craft workers or technicians, being responsive to changes in the education culture and context.

The implicit values and beliefs expressed in the professional knowledge landscape communicate clear expectations about practice. The official or sacred stories can influence the work of teachers, and can be observed at the level of classroom practice (as outlined by Clandinin and Connelly, 1995), and at the level of the school (Ball et al., 2012). These stories can exhibit values and beliefs but also expectations about practice. To behave in a way that is counter to these stories can be extremely challenging for teachers (for examples of this see Brooks, 2016).

This analysis would suggest then that teachers work in a context where the views of others, particularly of those with power and influence, can have a direct effect on their practice. It would be reasonable to assume then that influences from outside education (such as the growth of social media), which undoubtedly has an effect on education, could by extension affect how teachers work. But to explore this assertion, it is useful to understand how this interplay between context and identity works, how it can influence what we expect of teachers, and how we define expectations around their professional practice.

Professional practice

A broad view of teachers' professional practice sees the classroom as the main site of teachers' work and as the key location for their interactions with students and colleagues, whilst also recognising that some teachers' work takes place outside the classroom. The view of activist professionalism, as outlined by Sachs (2003), recognises that teachers may also be active in their subject communities, school communities and within other networks. Teachers may have perspectives on school policies, education policies, and research from both education and subject fields that can also influence their practice. Being a professional may spill over into their personal lives. Therefore engagement in social media, virtual spaces and networked communities are all potentially part of a teacher's professional practice.

This view of professionalism is that of a dynamic idea of 'being a teacher', a process that is being constantly made and remade in everyday actions and interactions. This perspective is summed up in Barnett's (2008) description of 'being a professional', drawing upon Heidegger's notion of being. This is not a static notion of being, as in to have arrived, but a dynamic sense of continual engagement and striving towards, as described up by Barnett (2008, p.193):

> 'Being' was a matter of how individuals are in the world, but it was, for Heidegger, also a matter of how an individual stood in time, backwards and forwards. Being has 'possibilities' and is always restlessly searching, working forwards, even if it knows not where or to what.

This view of professional practice would indeed suggest that online social spaces are likely to play an ever-increasing role in teachers' professional identity, due to their influence on the school context but also within society generally. However, such an assertion is insufficiently explicit. To explore what kinds of influences this can have, I now turn to recent work on teachers as curriculum makers, and explore how the way this perspective on teachers' work is conceptualised can offer some valuable challenges.

Teachers as curriculum makers

Detailed accounts of teachers' work are rare. The geography education community is fortunate to have two such accounts, both which focus on the subject department as a key locus of teachers work. The doctoral work of Steve Puttick (2015) and David Mitchell (2017) illuminates the complex network of influences, histories and traditions that can affect the dynamic of a department and the geography curriculum they enact with their students. The professional practice of the colleagues in question reveals forms of co-operation, negotiation and collaboration which draws upon individual expertise and shared understanding of goals and purposes (Puttick, 2016). It was his research in this area which led Mitchell (2016) to adopt his term of 'hyper-socialisation'. Mitchell (2016) describes what he calls the 'strikingly intense' (p.128) collaborative process of curriculum-making as follows:

> There is a carefully managed division of labour both in advance curriculum planning and at the point of classroom enactment. Each department followed the same curriculum-making process of
>
> (1) discuss and agree the overall content;
> (2) delegate the planning of learning objectives (LOs);
> (3) share LOs for feedback;
> (4) fix LOs and assignment guidelines for each unit;
> (5) delegate production of lesson materials;
> (6) share, amend and add to lesson materials on the department's intranet; and
> (7) use and rely on the intranet lesson plans and materials as they teach.
>
> Teachers download or 'pull off' lesson plans from 'the system' (department shared intranet), sometimes at the point of entry to the classroom. The teachers in each department work in close cooperation, even when there are some different opinions

about what and how to teach. LOs and assessments are agreed and adhered to, and through 'the system' teachers rely on one another to ensure curriculum materials are there when they need them. (p.128)

In his analysis of this process, Mitchell (2016) contends that this collaborative working is a response to the pressure that teachers find themselves under, a form of coping through harvesting resources, sharing them and co-operating with each other around their use. He highlights that this process is facilitated through broader networks and is social in nature, facilitated by online spaces (such as websites) where resources are made available. He also recognises that sharing is used as a way of reducing workload, and reducing the time taken up with planning and resourcing lessons:

> The 'social' nature of curriculum making is again striking by the level of sharing through wider communities, enabled by the Internet. This appears to be the enactment of Castells' (2004) globalisation through networks. The use of internet communities can, like the collaboration within the department, be interpreted as teachers' response to the need to 'perform' effectively and use time efficiently, coupled with the need to keep the geography curriculum 'relevant' (fun and engaging) for pupils as consumers. (Mitchell, 2016, p.128)

These accounts of the realities of geography teachers' work reveal how teachers are responding to the changing nature of the contexts that they work in. However, the accounts offered are centred around the (subject) department, and whilst Mitchell (2016) notes that the collection of resources involves other (outside) organisations such as the website of subject associations and media outlets, the question remains as to what extent this hyper-socialisation extends beyond the immediate professional circle of the teacher to the wider education community, and how in turn that will affect their identity. The argument developed so far is that influences on society and school contexts more generally will influence teachers' work, but online social spaces may behave differently. In order to understand this further, it is useful to draw upon a model of mobile learning developed in response to the proliferation of mobile devices used in learning environments, which outlines the parameters for engagement.

Socio-cultural ecological approach to mobile learning

Described as the mobile complex, the model below draws up a socio-cultural ecological approach as a conceptual framework which seeks to explain how the normalisation of mobile devices in everyday life affects opportunities for learning. The facets of this model can also be used to explore how opportunities for mobile and online resourcing (and in particular the curation of resources for teaching) can affect the cultural context of teachers' professional practice.

The model (see Figure 2.1) is made up of three interconnected components:

Agency: In the original model, young people are seen as displaying a new habitus of learning in which they see their environments as a potential resource for learning for which they are individually responsible. In the context of teacher professional practice, the environment, and in particular online communities, are also potential resources for learning, and the teacher, as the collector of such resources and as someone who engages in these networks, is agentic in how they seek, find and select – and indeed share – such resources.

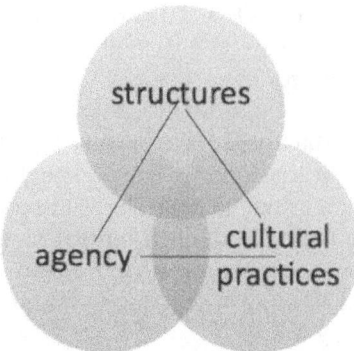

Figure 2.1 Key components of a socio-cultural ecological approach to mobile learning. From Pachler et al. (2010, p.4). Reprinted by permission.

Cultural practices: In the original model, mobile devices were mostly seen as useful for social interactions including communication and sharing, which Pachler et al. (2010) argue elevates everyday resources to achieve cultural significance. Similarly, for professional learning, everything potentially becomes an educational or classroom resource: teachers may curate resources from personal interactions, public websites, posts on social media; these are all potentially cultural signifiers for geographical content and concepts.

Structures: In the original model, mobile mass communication and the associated infrastructure sits alongside governed and structured formal educational systems. For teachers, the formal structures of textbooks and curricular materials are supplemented and perhaps supplanted by the resources obtained through similar mobile, social and technological infrastructures made easily available and easily transferable.

The examples given above relate to how teachers might use online spaces to supplement their classroom resources. But the adaptation of this model also reveals various socio-cultural dimensions to this phenomenon of hyper-socialisation, and therefore enables us to question more deeply the moral and ethical aspects of this element of professional practice.

The selection of any resource, including formal or endorsed textbooks, prompts a range of questions about the ownership, authenticity and accuracy of the resources being shared: to what extent have they been adequately and rigorously researched and verified? Who created the content? For what purpose? And what additional information was supplied with it that could be relevant? There are also issues around the selection and appropriation of the resources. To what extent are they representative of the phenomena, or partial in outlook? In addition, this model draws our attention to the social and cultural context in which this sharing takes places: who is sharing them in the first place? What networks do you need to belong to in order to have access to them? What degree of agency does the teacher demonstrate in their participation in this network? And what does this say about their identity construction?

The questions above are pertinent in discussions around professional practice, and it is not my intention to suggest either that teachers are somehow remiss in their professional judgement over the procurement of resources, or that online resources are more Machiavellian in intent or liable for inaccuracies than other resources. Indeed, the intention is to draw attention to the socio-cultural context that this online engagement happens within that introduces two important dimensions: that of agency and cultural practices.

Both agency and cultural practices are key factors in the identity work referred to previously: teachers as professionals enact their agency in a variety of ways, including in their participation in online spaces. But these actions take place within a cultural context (or professional knowledge landscape), and in the next section of this chapter I argue that online spaces have emerging cultural practices which can have different effects on teacher identity.

Online social spaces and teacher identity

Robson (2018) has developed a conceptual model based on his research into how teachers engage in online social spaces. Using the lens of professional identity, he has sought to illustrate the complexity and 'messiness' of the social realities of online spaces, and in doing so has conceptualised teacher identity work as being both performative and constructive. In his account he does not see these two dimensions, the performative and the constructive, as taking place separately but as linked with more active and passive forms of engagement. In other words, his work recognises the agency of individuals when they engage with online social contexts in managing the perceptions of themselves and their peers, albeit in idealised forms. These formations are taken within a context of a shared understanding of an 'ideal identity' which forms the basis of a social structure (after King, 2005). Robson (2018) argues that 'These embedded ideals were then reproduced through performance ensuring the ongoing constructive power of the online social structure' (pp.446–447), which he terms 'professional normalisation'. He concludes: 'Shared understandings of professionalism and embedded ideal identity positions were internalised, aspired to, performed and discursively reproduced' (Robson, 2018, p.447). Robson's (2018) model reflects the complexity of engaging in online social spaces and the struggle that teachers can experience when they engage in open online social networks.

As a concrete example, EduTwitter is the term given to a number of users of the social media platform Twitter who converse about education. In a TES article, Bisphan (2018) outlines his own experience of having sent a tweet with a hyperbolic statement about teachers who vote Conservative:

> It was designed for professionals to read and I made assumptions about the ability to make inferences about my frustration. However, it was twisted and taken out of context and before I knew it, I was added into threads in which a small and cult-like group of fellow teachers (sometimes from faceless accounts) attacked me. It was sobering to realise how desperate ideologues are to stamp a claim on the social media landscape.

He attributes the backlash to his tweet to a number of groups of people on Twitter who subscribe to particular ideological viewpoints about education and use the platform to 'advocate and impose their own ideology on other professionals' (Bisphan, 2018). This example shows how the social space of an online platform is used to construct identities, idealised identities, for individual teachers. Periodically debates on Twitter spiral into personal attacks; the summer of 2019 for instance was the summer of '#listgate' when one Twitter user generated a list of who they considered to be useful people to follow, and was then opened to a barrage of accusations of racism and other forms of partiality due to the range of people on the list, which were then staunchly defended by another group of Twitter users. The criticisms may have been fair, the aggressive and personal nature of the attacks less so. The issue here is that whilst many claim Twitter and other social media outlets to be valuable forms of professional development, they are also public spaces where comments are viewed, commented upon and are constitutive of an online social structure. Social online space and social media are made up of embedded ideals of teachers and teacher behaviour, and processes through which individual agency is enacted, monitored and to some extent controlled and constructed.

This is further emphasised in cases where teacher social media accounts or online social presences become known beyond the teacher community and are shared by parents, colleagues and pupils. In these spaces, the idealised form of teacher identity is strongly enforced (see Hepburn, 2018). As Robson (2018) highlights, these spaces can then become highly constructive of teacher identities, with teachers self-monitoring their public image, and deliberately concealing private aspects of their identity (particularly around what might be perceived as sensitive issues such as religion or sexuality).

To return to the model in Figure 2.1, the social media space and structure employs particular cultural practices which can limit individual agency. In this sense, EduTwitter, and other forms of social media, are not benign spaces for teacher professional development as claimed by some (Caron, 2011), but become spaces of performance. Or as Robson (2018, p.447) notes:

> As teachers increasingly turn towards online social spaces for peer-to-peer interaction … there is a need to move beyond instrumental discourses relating to teacher development and problematise the spaces themselves in a way that reflects the messy social realities teachers engage in.

Curriculum curation

Robson's (2018) work focusses mainly on the implications for teacher identity with regards to online social media engagement. But his observations, and the relationships illustrated in the mobile learning model (see Figure 2.1), can also be used to understand the role that online spaces can play for teachers as curriculum makers. Mitchell's account of geography teachers as curriculum makers shows that the curation of resources came from a variety of places: online news websites, formal sources of teacher resources such as the websites of professional associations, social media. In some cases the sources of these resources could be considered as authoritative: official government sources, NGOs, or respected international organisations such as UNESCO. Charities also use online spaces as a way of promoting their work, and this can involve informative videos, fact-sheets and case studies – all of which can be useful for the geography teacher. Some sources are less formal: sharing from other teachers, colleagues, friends, or personal accounts, anecdotes, photographs. And it would also be true to say that before the proliferation of the internet, teachers, and particularly geography teachers, would collect geographical resources in a variety of places. The internet has increased the volume of potential resources, whilst also increasing the ease with which they can be collated, shared and adapted for classroom use (Puttick, 2021; Roberts, 2021).

As with the selection of any teaching resource, there are a number of considerations that need to be taken into account: appropriateness of the information, how it is presented, how it is to be used in the lesson, and the extent to which it represents the learning intentions. Related to these considerations are also concerns about the resource itself: the accuracy of the information, the source and the extent to which it is reliable and appropriate. Whilst I would contend that most teachers are aware and diligent of such considerations, there is a temptation with the immediacy and availability of some online resources to accept and integrate them without question.

Here it is useful again to return to the model of mobile learning: the structures and availability of resources, and the immediacy of use, make up the processes with which they are available, but are also influenced by softer cultural practices around sharing, which in turn will influence the degree of agency a teacher exercises. For example, a teacher sharing a resource on Twitter, perhaps with an accompanying tweet about how useful they found it in their own practice, is likely to influence how another teacher responds to that resource, how they interrogate it and consider the reliability and the veracity of information contained therein. The more respected the teacher doing the original sharing (particularly if they have a large following on Twitter) the more likely it is that their endorsement will carry extra weight. This is not to suggest that resource-sharing of this nature is not a valuable endeavour. The intention is to call attention to the social and cultural context within which this encounter takes place, and the potential impact on teacher agency.

Bearing this in mind, I contend that it is useful to view the act of harvesting resources from a range of online sources as a form of *curation*. Traditionally used in relation to art or museum exhibitions, the act of curation is the drawing together of artefacts, artists or content for their intended purpose. This act requires separate actions of selection, organisation and presentation: each requiring a different set of questions about source, purpose and use. The key dimension of curation is that it is done by an expert: someone knowledgeable in the field, who uses their knowledge to make value judgements of inclusion and exclusion, but also in the presentation, ordering and display of those items. This metaphor works well for resources for learning: the teacher needs to see themselves in the position of expert, enabled to make decisions about selection, but also drawing upon their knowledge of their students, their prior learning and their learning goals as to how that information should be presented, questioned and used to develop learning in the classroom. It combines, therefore, the idea of agency, structure and cultural practices, enabling teachers to act with a high degree of professionalism.

Viewing curation as a key component of teacher professionalism is, I would argue, advantageous from the perspective of teacher identity. It recognises the expert knowledge of the geography teacher, their adaptive expertise, and their professional judgement in the light of what resources are useful, for which students, and for what purpose. It also recognises that their engagement in online social spaces is from the position of a professional: not just a 'follower' but of someone who has agency and who can exercise judgement – a colleague. In both cases then, the notion of curation can nourish teacher identity (and particularly their subject identity) as it enforces the importance of professional values, and what it is that teachers consider to be important.

References

Ball, S. and Goodson, I., 1985. *Teachers' Lives and Careers.* London: Falmer Press.

Ball, S.J., Maguire, M. and Braun, A., 2012. *How Schools Do Policy: Policy Enactments in Secondary Schools.* London: Routledge.

Barnett, R., 2008. Critical professionalism in age of supercomplexity. In B. Cunningham (ed.), *Exploring Professionalism.* London: Institute of Education, University of London, pp.190–208.

Bisphan, J., 2018. Don't get sucked into trad vs prof on EduTwitter. *Times Educational Supplement*, 22 October. Available at www.tes.com/news/dont-get-sucked-trad-vs-prog-edutwitter [Accessed on 19 May 2020].

Brooks, C., 2016. *Teacher Subject Identity in Professional Practice: Teaching with a Professional Compass*. London: Routledge.

Caron, S.W., 2011. Using Twitter for professional development. *Education World*, 2 March. Available at www.educationworld.com/a_tech/using-twitter-for-professional-development.shtml [Accessed on 20 May 2020].

Castell, M., 2004. *The Network Society: A Cross-Cultural Perspective*. Cheltenham: Edward Elgar.

Clandinin, D.J. and Connelly, F.M., 1995. *Teachers' Professional Knowledge Landscapes*. New York: Teachers College Press.

Day, C. and Gu, Q., 2010. *The New Lives of Teachers*. Abingdon: Routledge.

Day, C., Sammons, P., Stobart, G., Kington, A. and Gu, Q., 2007. *Teachers Matter: Connecting Work, Lives and Effectiveness*. Maidenhead: Open University/McGraw-Hill.

Hepburn, H., 2018. Social media 'catastrophic for teachers. *Times Educational Supplement*, 12 June. Available at: www.tes.com/news/social-media-catastrophic-teachers [Accessed on 19 May 2020].

King, A., 2005. Structure and agency. In A. Harrington (ed.), *Modern Social Theory: An Introduction*. Oxford: Oxford University Press, pp.215–232.

Mitchell, D., 2016. Geography teachers and curriculum making in 'changing times'. *International Research in Geographical and Environmental Education*, 25, pp.121–133.

Mitchell, D., 2017. Geography Curriculum Making in Changing Times. Unpublished PhD thesis, University College London.

Pachler, N., Cook, J. and Bachmair, B., 2010. Appropriation of mobile cultural resources for learning. *International Journal of Mobile and Blended Learning*, 2, pp.1–21.

Puttick, S., 2015. Geography Teachers' Subject Knowledge: An Ethnographic Study of Three Secondary School Geography Departments. Unpublished DPhil thesis, University of Oxford.

Puttick, S., 2016. An analysis of individual and departmental geographical stories, and their role in sustaining teachers. *International Research in Geographical and Environmental Education*, 25, pp.134–150.

Puttick, S., 2021. Digital technologies and their roles in knowledge recontextualisation and curriculum making. In N. Walshe and G. Healy (eds), *Geography Education in the Digital World*. London: Routledge, pp.17–25.

Roberts, M., 2021. Geographical sources in the digital world: Disinformation, representation and reliability. In N. Walshe and G. Healy (eds), *Geography Education in the Digital World*. London: Routledge, pp.53–64.

Robson, J., 2018. Performance, structure and ideal identity: Reconceptualising teachers' engagement in online social spaces. *British Journal of Educational Technology*, 49, pp.439–450.

Sachs, J., 2003. *The Activist Teaching Profession*. Maidenhead: Open University Press.

Sammons, S.P., Day, C., Kington, A., Gu, Q., Stobart, G. and Smees, R., 2007. Exploring variations in teachers' work, lives and their effects on pupils: Key findings and implications from a longitudinal mixed-method study. *British Educational Research Journal*, 33, pp.681–701.

Chapter 3

Digital technologies and their roles in knowledge recontextualisation and curriculum making

Steve Puttick

Introduction

I am currently editing parts of this chapter and re-reading papers discussing the 'digital turn' in geography (Ash et al., 2018) in New Delhi, using the same cloud service that I normally access from the UK, and seeing the same information through browsers and other programs (reference management software, a word processor and a pdf viewer). The assemblage of table, laptop, WiFi router and software constructs a reading and writing 'code/space' (Dodge and Kitchin, 2005, p.172). Similarly, on my flight to Delhi, my reading and writing code/space was in some ways quite different (35,000 feet higher, and 570 miles per hour faster) but in other ways very similar, including information mediated through the same software, reliant on the same code and hardware. There is something very convenient and exciting about the immediacy of this digital connection. But this does not mean that we should be lured into inferring that the codified knowledges that I am 'freely' accessing in India are, therefore, freely accessible globally. Teachers' access to the digital technologies through which this information is mediated is not evenly distributed. Nor are the vast quantities of information that are digitally mediated representative or evenly distributed (Graham et al., 2014), either in terms of their origins of production or foci of attention:

> These new digital geographies (both social and economic) are by no means technologically determined. Rather, the way in which places and people become 'wired' (or remain 'unwired') still depends upon historically layered patterns of financial constraint and cultural and social variation. The geographic and technological evolution of this digital infrastructure can therefore be understood as a process of social construction of new (and often personal) digital geographies. These new geographies are both immensely empowering (for the people and places able to construct and consume them) and potentially overpowering as institutional and state forces are able to better harness information with growing personal and spatial specificity. (Zook et al., 2004, p.156)

The small text on the top of my browser tells me through which institution I am accessing the material. There is a substantial charge the institution pays the multinational publishing company for access, funded in part by the fees from the international students that I am in India to interview, but who in their current position as school teachers in India (and the same is true for most school teachers globally) cannot access beyond the abstract. The particular paper I am currently reading is also only available in English, ruling out access to the 80% of the world's population who cannot read English.

I mentioned above a range of digital technologies used to write a chapter (reference software, pdf viewers, word processor and cloud storage services), and this is a shorter list than the range of digital technologies teachers often use to construct and resource school geography lessons. There is much that needs to be explored around the roles that other digital technologies play in teachers' curriculum making. For example, the epistemological, representational and pedagogical implications of PowerPoint slides. This chapter critically examines the origins, journeys, transactions and recontextualising processes (Bernstein 2000) through which information travels into school classrooms mediated by digital technologies.

The 'digital turn' in geography

There are big claims made about the roles of digital technologies – and more broadly, of 'computers' – in and on geography. Sui and Morrill (2004) argue the computer is the single most significant technological innovation to have affected geography, 'drastically transform [ing] both geography as an academic discipline and the geography of the world' (p.82). The multiple intersecting influences through, by and of the digital, create change in different directions, and so the account of the 'digital turn' in geography described by Ash et al. (2018; 2019) surveys 'geographies produced *through*, produced *by* and *of* the digital' (2018, p.25).

The digital turn might be seen as following previous 'turns' in the discipline (Taylor, 2009), including the cultural turn of the 1990s (Barnett, 2002) and the affective turn of the 2000s (Leys, 2011). Initially, work in the digital turn 'concentrated on how ICTs, and the internet in particular, were transforming economic, cultural, social, and political geographies' (Ash et al., 2018, p.30). As this work has developed, the spatial metaphors of infrastructure through which digital technologies are imagined and described have foregrounded the importance of geography in understanding and researching them: 'metaphors of information superhighways and wired cities are useful in imagining a world in which data is created, shared, accessed, and cross-checked in historically unprecedented volumes' (Zook et al., 2004, p.156). Ash et al. (2018) define the digital as

> material technologies characterized by binary computing architectures; the genre of socio-techno-cultural productions, artefacts, and orderings of everyday life that result from our spatial engagement with digital mediums; and the logics that both structure these ordering practices as well as their effects ... [and] digital discourses which actively promote, enable, secure, and materially sustain the increasing reach of digital technologies. (p.26)

They argue that there is a 'need to more fully consider how the digital inflects geography's many subfields and mediates how geographical knowledge is produced' (Ash et al., 2018, p.27). There are significant questions about how the digital inflects the geography of education, and particularly teachers' work in curriculum making and knowledge recontextualisation. The threefold categorisation of the relationship between geography and the digital (geographies produced *through*, produced *by*, and produced *of* the digital) might be traced through geographies of teachers' practice: the space-times of teachers' journeys for knowledge produced *through* the digital; the production, representation and mediation *by* the digital of the geographical knowledge used for and accessed in school geography lessons; and – as an area of study to be taught to and explored by students – the geographies *of* the digital. Digital technologies transform geography education and teachers' practices in curriculum making in significant ways (including accessing information online and ICT presentation tools), but these same technologies are also (or at least have the potential to be) repurposed, reshaped and remade by geography education. This might include geography education research developing improved understandings of teachers' knowledge recontextualisation in the context of digital technologies leading to improved practice, and teachers engaging young people with more critical understandings of digital technologies that empower them to rethink and more actively shape the way they use them. While digital technologies are significant to the broader discipline, and are actively engaged with in students' and teachers' daily lives, they are largely absent from school geography curricula globally. This is a missed opportunity, denying teachers and students key geographical lenses through which they might more critically understand and act on the pervasive digital geographies they are almost constantly participating in.

An important aspect of research on digital geographies has been the uncovering of power relations and the ways in which power relations transform and are transformed through new digital technologies. Some of the relations are new and involve novel spatialities (such as the creation of code/spaces). However, rather than understanding digital geographies as something completely novel, they:

> are best understood through the lenses of extant as well as emerging fields of geographic inquiry ... how digital phenomena, practices, and presences inflect and reconfigure geographical thinking about and approaches to questions of epistemology and knowledge production, space and spatiality, methods and methodologies, culture, the economy, and politics. (Ash et al., 2018, p.27)

As such, 'the rise of digital content comprises new forms of evidence with which to approach long-standing geographical concerns; and digital presences and praxes are provoking new questions and opening up new lines of geographical enquiry' (Ash et al., 2018, p.27). The digital turn then begins to 'capture the ways in which there has been a demonstrably marked turn to the digital as both object and subject of geographical inquiry, and to signal the ways in which the digital has pervasively inflected geographic thought, scholarship, and practice' (p.25). One dimension of the continuity with the past is illustrated with the example of the teacher's journey for information online mirroring their previous journeys, such as those to physical books.

Space-times of teachers' journeys for information

The digital turn has presented significant opportunities for geography teachers in terms of the timeliness, quantity and variety of information they are able to access. The following description of teachers' journeys for information considers some of the ways in which these journeys are similar to previous (non-digital) journeys, and also some of the ways in which they are very different. The illustrations are drawn from doctoral research: an ethnographic study of three secondary school geography departments, focusing on geography teachers' subject knowledge. Further contextual information about the schools, departments and teachers are explored in more detail elsewhere (Puttick, 2015; 2016; 2017).

Teachers describe sources of knowledge they use in their lessons in mainly spatial and temporal terms, particularly in terms of them 'going to ... and getting'. For example, using a nautical metaphor, Gemma (a geography teacher interviewed during the research) described how her 'first port of call – like most people – is the internet'. 'Port of call' is an interesting term; 'it speaks of journeys begun, of loading up, and setting out on further journeys' (Puttick, 2014, p.114). Search engine algorithms, which are increasingly personalised and responsive to past behaviours and choices, play a significant role in the visibility – and associated legitimacy – of different sources of information available online (Puttick, 2017), and in turn influence geography teachers' curriculum making (Mitchell, 2020). It will be increasingly important for geography education research to engage with questions over the accessibility and visibility of different sources. Information that is available, whether through textbooks and other published resources, and particularly through 'traditional mass media, tend to reinforce the already visible and powerful at the expense of minority or oppositional perspectives' (Graham et al., 2014, p.747). The following example from another teacher highlights the continuities between these very different physical and online journeys for information:

> ... now all I've got to do is go online and I can find that diagram, in an electronic form, and I just copy it straight onto a page, and I don't have to do what I used to do in the past, which is have to photocopy the page, cut out the bit that I want, actually glue that onto a page, and then, y'know ... your Rostow's model, and your – standard things that have stood the test of time – it's just where I get the model or the diagram from now. I don't have to go to the book. (Ruth, interview)

Here, Ruth's journey for information has changed from a physical journey across her classroom to a digital one, but the purpose is identical: a copy of the same model. The ways in which information is presented by search engine results are different from Ruth's previous journey – for example, across the classroom to open a cupboard and take out physical copies of textbooks. This seems to be a fascinating development: the *potential* sources of available information have increased exponentially. Yet, the actual sources of information are remarkably constant. There are opportunities for geography teachers to diversify the range, origin and scope of information accessed, and there are opportunities for geography education researchers to critically explore the knowledge politics of the school subject in the digital world, drawing more broadly on digital geographies research. Aspects of existing formal curricula present some of these opportunities already. For example, the aim of the Key Stage 3 National Curriculum in England is to introduce young people to human geographical features of the world, which necessarily includes some understanding of the ways in which these human geographies are constructed through and mediated by digital technologies. One reason why there has been little engagement in the school subject with digital geographies may be associated with the power of examination specifications (Puttick, 2015). The next iteration of the Geography National Curriculum and subsequent examination specifications would be a great chance to engage more explicitly with this area of the subject.

Existing digital geographies research suggests there is little to indicate that 'the digital' results in greater equity and less reproduction of dominant perspectives. There are obvious continuities in terms of knowledge production and the power relations associated with this production: 'knowledge and codified information about the social, economic, and political contexts to our lives are always produced under conditions of power' (Graham et al., 2014, p.747). The conditions of power seem to be relatively stable and insulated from change that might be affected by digital technologies. The difference is mainly felt in speed and scale: 'the internet and mobile phone have changed the amount, diversity, speed and geographical scale of communication beyond recognition over the past 40 years' (Schwanen, 2019, p.60).

A further and contrasting example of a different teacher's journey for information is presented in the next section, highlighting some of the ways in which the amount, diversity, speed and scale of information that is *potentially* available for his lessons represents a transformation from that which was previously limited to the cupboard at the back of the classroom.

Intensification of teachers' work and curriculum making

The boundaries between work and home are increasingly blurred (Graham and Anwar, 2019), and teachers' engagement with the digital technologies through which they interact with their department and curriculum making extends beyond the physical boundaries of the school. All teachers in the study described above (Puttick, 2017) accessed news sources online. Some occasionally accessed articles only when teaching a new topic, whereas others curated articles from news apps on their smartphones every morning before school. In Ruth's case, she represents a slightly clearer separation of work and home, in that lesson planning is all described in terms of what she does at her desktop computer in school. Most other teachers described more pervasive 'always on' mobile lives (Schwanen, 2019). For example, one Head of Department described the ways in which they used their smartphone at the breakfast table each morning, checking emails, scanning Twitter feeds, and choosing articles from the day's news that would be relevant to that day's lessons. These practices have implications for how we understand teachers' professional knowledge and the kinds of demands that are placed on it, including the ways in which this embodied knowledge is used to make very rapid decisions in selecting or rejecting articles.

The kinds of information about an event that might have happened on a different continent while the teacher and students were sleeping represents – temporally at least – a disjunction with previously available sources of information. Shifts in the temporal demands on teachers – including the rapid, ongoing judgements about sources of information in the construction and reconstruction of their subject knowledge – is an important area for research on teachers' recontextualisation of knowledge and curriculum making to explore. Possibly, such research might consider observing what teachers do on their computers (and smartphones / tablets), investigating the ways in which teachers make decisions about what articles to include or exclude. In particular, this research might focus on issues of legitimacy and trust in the use of websites, as considered by Roberts (2021), and the selection of particular articles.

A tsunami of information

The discussion now moves from teachers' journeys for information and in to school classrooms, providing one example of the 'live' judgements teachers might make about information accessed through media, including digital technologies.

The teacher (Hugh) began one lesson with a Year 7 class by telling them he was slightly concerned about their use of the internet. He warned them of a tendency to 'sit and act like mirrors', in which they simply reflect the information they see on the screen. He continued, introducing the lesson:

HUGH: I've got a whole pile of magazines there – it's a bit of a risk, but [noise from a student], are you a gambler Jack? I've seen what you get up to at break time! There are *Geographical* magazines there … Just flick through – I want to get you looking for information; how do you gather it? I love the images – I think some of the photos are absolutely phenomenal. Take some time to just look at them. That's where some of my love for geography actually comes from. Jack, why are you laughing?

JACK: I've seen a photo of a goat and it looks like it's got a beard.

HUGH: Right, where is it?

JACK: In the Rocky Mountains.

HUGH: So there we go – Jack's already found something he's interested in – the bearded goat … I am going to let you use the iPads, I am going to let you use the computers – we're gathering information.

(Lesson observation)

After this introduction and the exchange between Hugh and Jack, the students flicked through the *Geographical* magazines. One group became interested in an image of a 'tsunami'. It is a dramatic image that has been used in a number of different contexts. They were looking at the *Geographical* magazine, where it takes up the top third of the page, above the headline 'US west coast tsunami threat upgraded'. The image shows the downtown area of a large US city on the west coast, including tall buildings, transport hubs, infrastructure, industry and housing. Towering high above this modern coastal urban environment is the spray from an enormous wave. The composition of the image emphasises the height of the wave: the lower half of the image is taken up with the city, and the whole of the top half is spray from the wave. On the right-hand side, a part of the unbroken crest of the wave is also visible, towering many times higher than the highest skyscrapers in the city below.

The group of students then used their tablets to search online for 'tsunamis'. On the first line of their Google search they found a similarly striking image of a modern coastal city with high tower blocks, housing and infrastructure all of which is about to be engulfed by another enormous wave, the crest of which reaches over the top of the highest building in the city (which is 100m+ high). Again, this image is highly recognisable and has been used on a number of teaching resources about tsunamis that are widely available online.

When Hugh saw the tsunami images, he was shocked. His response was to quickly source, through his own Google search, a different photograph of a tsunami that contrasted strongly with the images the students had found. In his photograph, taken facing out from the shore, a tsunami wave is approaching and a section of a coastal town or suburb is shown with approximately a dozen one or two-storey houses, and in front of them a row of trees. The tsunami wave reaches up to approximately the first quarter of the trees that it is passing. At its tallest point, the wave might be five metres high.

The *Geographical* article is published under an image of a city being engulfed by a huge wave. So huge that it is many times higher than the highest known tsunami wave. The text below the image emphasises wave height: 'In 1964, the Alaksan earthquake, the second-largest ever recorded, triggered tsunamis as high as 52 metres. The new research suggests that a more extensive earthquake may create even bigger tsunamis' (Yap, 2009, p.5). The 'new research' cited is an article in *Quaternary Science Reviews* (Shennan et al., 2009). The image of the tall wave, combined with the text about wave heights and predicted increases in the size of tsunamis is highly dramatic. Shennan et al. (2009) argue that their evidence, which indicates widespread rupturing of the Yakutat microplate (a relatively small tectonic plate in southern Alaska), has 'significant implications for the tsunami potential and seismic hazards of the Gulf of Alaska' (p.12). They specifically mention wave height only to suggest that, even when earthquakes are larger, the specific conditions in these areas (in particular, large areas of uplift, and a relatively shallow sea floor) result in *similar* wave heights. The highest wave recorded in 1964 was confined to one inlet, and was caused by a localised landslide, and as such was a very specific event. Shennan, Bruhn and Plafker cite Nanayama et al. (2003) in support of their argument that the multi-segment earthquakes (related to the widespread rupturing of the plate) may produce a tsunami 'with greater wavelength that penetrates farther inland, even though the height of the wave at the coast may be similar' (Shennan et al., 2009, p.12). The *Geographical* article's representation of this research might be seen as an example of journalistic sensationalism. What is interesting here is the way in which Hugh responded with particular kinds of recontextualising principles to source additional information (the town or suburban image described above) and engage the students in a dialogue, and all of this in a very short space of time.

I wrote the above discussion of the image whilst sitting at my desk. I spent two hours or more reading the journal articles mentioned, checking terms and concepts I was unfamiliar with, and a further hour writing the summary paragraph I have presented here. I did not have responsibility for 30 Year 7 students in my office. Hugh's critique of the photograph with the students took less than two minutes. Like the students, he did a Google image search, and then placed a photograph of a real tsunami alongside the altered images they had found. He asked the students questions about the images, prompting them to compare and contrast the sources: how high is that wave up those trees? How high are skyscrapers? Is this a real wave? The discussion around the images was dichotomous; the students' task was to classify them as right/wrong, and real/fake. Another interesting question raised through this episode is about teachers' responsibilities for information drawn on in their recontextualisation of knowledge for school geography lessons. This example illustrates the kinds of dynamic decisions that teachers make, and the active ways in which their professional knowledge is enacted. Digital technologies do not create completely new demands on teachers' curriculum making, but an intensification of these demands by speeding-up access to and massively expanding the quantity of different sources of information. This example of critically analysing the ways in which a tsunami is represented, including the sensationalising of associated physical processes, might be usefully replicated across other areas of the subject, in particular for research to critically explore the ways in which topics such as migration and climate change are represented and mediated through digital technologies. One significant aim of this work is to support teachers' curriculum making in ways that maximise the opportunities offered by digital technologies that are sensitive to the tendency of neoliberal marketisation forces to favour sensationalised representations.

Conclusions and implications for practice

Digital geographies present many interesting opportunities for geography teachers, including as a substantive area of study at the cutting-edge of the discipline, and as a *potentially* diverse and up-to-date source of information. Curriculum making is *potentially* enriched enormously through critical engagement with digital geographies. I have argued that recent research in digital geographies sheds light on important areas of teachers' practices and the geographies of education. I have also argued that this research raises important critical questions for geography education to explore. Drawing on ethnographic research with secondary school geography departments, I have offered examples of the ways in which 'new' digital geographies maintain significant continuities with previous practices. There is a danger that the ubiquity of 'open' and 'freely available' digital information creates the illusion that we are making geography curricula more open. Evidence from the wider body of digital geographies research suggests this is highly unlikely, particularly because of the ways in which language, politics and power relations are reproduced and even exacerbated through the intensification of speed and quantity. Another aspect of intensification within the broader context of teachers' recontextualisation and curriculum making in a digital world relates to blurring work/life distinctions. The ethnographic accounts of teachers describing their curation of information (primarily through social media feeds and online news) on smartphones at the breakfast table before school is an important dimension – of some teachers' work enabled by the digital – to consider alongside questions about *what* is accessed. For example, asking: how is teaching imagined, experienced and transformed through digital technologies? How are new technologies transforming the capacities of teachers in their curriculum making? In what ways can initial teacher education and ongoing professional development best support teachers' knowledge practices for the task of curriculum making in a digital age? These questions include conceptual, methodological and empirical areas for further research, and offer great potential for collaborations between geography teachers, geography education researchers, and academic geographers (and, obviously, these are not necessarily distinct categories).

One further implication for those involved in curriculum making is the potential for greater engagement with digital geographies as an area of study, both as a distinct topic and as an integral dimension of existing areas of study (Ash et al., 2018). In the latter case, the curriculum challenge focuses on 'how an engagement with the digital develops our collective understandings of cities and development, as well as health, politics, economy, society, culture, and the environment' (Ash et al., 2018, p.35). This might include: teachers enriching aspects of Key Stage 2 and 3 curricula with conceptual resources drawn from digital geographies, such as introducing some of the ways in which geographical knowledge is constructed and mediated including analysing representations of human and physical environments in computer games played by young people (for an example of work in this particular area that might be usefully recontextualised for school geography, see Ash et al., 2009). Finally, future revisions of the geography national curriculum and exam specifications might reflect on the ways in which critical engagement with digital geographies has the potential to enrich the subject, which in the short term might include rearticulating some aspects of existing specifications to make this area of the discipline explicit to teachers.

References

Ash, J., Kitchin, R. and Leszczynski, A., 2018. Digital turn, digital geographies? *Progress in Human Geography*, 42 (1), pp.25–43.
Ash, J., Kitchin, R. and Leszczynski, A., 2019. *Digital Geographies*. London: Sage.
Ash, J., Romanillos, J.L. and Trigg, M., 2009. Videogames, visuality and screens: Reconstructing the Amazon in physical geographical knowledge. *Area*, 41 (4), pp.464–747.
Barnett, C., 2002. The cultural turn: Fashion or progress in human geography? *Antipode*, 30 (4), pp.379–394.
Bernstein, B., 2000. *Pedagogy, Symbolic Control and Identity: Theory, Research, Critique*. London: Rowman & Littlefield.
Cresswell, T., 2011. Mobilities I: Catching up. *Progress in Human Geography*, 35 (4), pp.550–558.
Dodge, M. and Kitchin, R., 2005. Code and the transduction of space. *Annals of the Association of American Geographers*, 95 (1), pp.162–180.

Graham, M. and Anwar, M.A., 2019. Labour. In J. Ash, R. Kitchin and A. Leszczynski (eds), *Digital Geographies*. London: Sage, Chapter 16.

Graham, M., Hogan, B., Straumann, R. and Medhat, A., 2014. Uneven geographies of user-generated information: Patterns of increasing informational poverty. *Annals of the Association of American Geographers*, 104 (4), pp.746–764.

Leys, R., 2011. The turn to affect: A critique. *Critical Inquiry*, 37 (3), pp.434–472.

Mitchell, D., 2020. *Hyper-Socialised: How Teachers Enact the Geography Curriculum in Late Capitalism*. London: Routledge.

Nanayama, F., Satake, K., Furukawa, R., Shimokawa, K., Atwater, B.F., Shingeno, K. and Yamaki, S., 2003. Unusually large earthquakes inferred from tsunami deposits along the Kuril trench. *Nature*, 424 (660), pp.660–663.

Puttick, S., 2014. Space-times of teachers' journeys for knowledge. *Teaching Geography*, 39 (3), pp.114–115.

Puttick, S., 2015. Chief examiners as prophet and priest: Relations between examination boards and school subjects, and possible implications for knowledge. *The Curriculum Journal*, 26 (3), pp.468–487.

Puttick, S., 2016. An analysis of individual and departmental geographical stories, and their role in sustaining teachers. *International Research in Geographical and Environmental Education*, 25 (2), pp.134–150.

Puttick, S., 2017. 'You'll see that everywhere': Institutional isomorphism in secondary school subject departments. *School Leadership & Management*, 37(1–2), pp.61–79.

Roberts, M., 2021. Geographical sources in the digital world: disinformation, representation and reliability. In N. Walshe and G. Healy (eds), *Geography Education in the Digital World*. London: Routledge, pp.53–64.

Schwanen, T., 2019. Mobilities. In J. Ash, R. Kitchin and A. Leszczynski (eds), *Digital Geographies*. London: Sage, Chapter 6.

Shennan, I., Bruhn, R. and Plafker, G., 2009. Multi-segment earthquakes and tsunami potential of the Aleutian megathrust. *Quaternary Science Reviews*, 28 (1–2), pp.7–13.

Sorensen, P., Twidle, J. and Childs, A., 2014. Collaborative approaches in initial teacher education: Lessons from approaches to developing student teachers' use of the internet in science teaching. *Teacher Development*, 18 (1), pp.107–123.

Sui, D. and Morrill, R., 2004. Computers and geography: From automated geography to digital earth. In S.D. Brunn, S.L. Cutter and J.W. Harrington (eds), *Geography and Technology*. Dordrecht: Springer, pp.81–108.

Taylor, C., 2009. Towards a geography of education. *Oxford Review of Education*, 35 (5), pp.651–669.

Yap, R., 2009. US west coast tsunami threat upgraded. *Geographical*, October, p.5.

Zook, M., Dodge, M., Aoyama, Y. and Townsend, A., 2004. New digital geographies: Information, communication, and place. In S.D. Brunn, S.L. Cutter and J.W. Harrington (eds), *Geography and Technology*. Dordrecht: Springer, pp.155–176.

Chapter 4

Navigating the theory-practice divide
Developing trainee teacher pedagogical content knowledge through 360-degree immersive experiences

Nicola Walshe, Paul Driver and Mandy-Jane Keenoy

Introduction

The significance of teacher and, thereby, trainee teacher subject knowledge is currently under scrutiny with a commitment to subject-knowledge development and subject-specific pedagogy emerging from the government through various policy documents (e.g. the Carter Review; Carter, 2015). However, within a complex initial teacher education (ITE) landscape which includes a range of school-centred ITE pathways and programmes, there remains a concern that trainee teachers lack support in developing their subject expertise (pedagogical or otherwise). Brooks (2017) argues that trainees need to be supported to develop a subject-specific approach to pedagogy, yet many remain insufficiently supported to understand the complex relationship between subject and pedagogy and, therefore, are unable to articulate the relationship between the two through either lesson observations or their own teaching practice.

Video technologies are posited as having the potential to provide opportunities for trainee observation and reflection, for example on the relationship between subject and pedagogy, providing an 'unbiased authenticity to reflective dialogue' (Calandra and Brantley-Dias, 2006, p.137). Despite this, research has found that the use of video with beginning and experienced teachers is often limited to providing examples of expert practice (e.g. Borko et al., 2011), with little scaffolding to develop meaningful reflection. More recently, 360-degree video has been shown to have potential to help trainee teachers relate theoretical learning to classroom practices (Ibrahim-Didi, 2015; Walshe and Driver, 2018), making progress towards bridging the gap between theory and practice (Abell and Cennamo, 2004). However, although the use of video within ITE is increasing (Tripp and Rich, 2012), there is limited research on how best to utilise the unique capabilities of 360-degree immersive video. In response to this, this chapter explores how 360-degree video, particularly what we term 360-degree immersive experiences, might be used to support trainee teachers to develop a clearer understanding of subject and pedagogical knowledge for teaching. In particular, it focuses on use of these experiences for scaffolding trainee teachers' lesson observation as a precursor to exploring pedagogical content knowledge (PCK; Shulman, 1986).

A lack of noticing

Over the last ten years, there has been a change of focus to school-centred training in teacher preparation programmes (Murray and Mutton, 2016), reflecting an international 'practical turn in teacher education' (Mattsson et al., 2011, p.17) that is represented by an increased focus on the amount of in-school experience within ITE. This has resulted in trainee teachers spending more time observing practice in school; however, it has been argued that it is not always clear what they learn from these experiences (Brophy, 2004). For example, research has highlighted trainees in exposure to initial classroom lessons are uncertain and tend to focus on irrelevant, superficial events, due to a lack of experience and pedagogical understanding (e.g. Sherin and van Es, 2005; Stürmer et al., 2013). This could be attributed to the fact that, within a given classroom context, there are at any one time a multitude of interactions taking place, decisions being made and outcomes emerging. Indeed, Star et al. (2011) suggest that such sensory overload in the classroom environment is not conducive for trainee teachers to distinguish the salient details, let alone begin to understand the processes taking place. Moreover, the relative inexperience of a trainee teacher often means they lack an epistemological framework which would enable understanding and identification of significant classroom events (Scherr and Hammer, 2009; Luna and Sherin, 2017). Because of this, Sherin and van Es (2005, p.489) suggest the need to provide opportunities and structures to scaffold trainee teachers' 'ability to notice'. This view is supported by Grossman and McDonald (2008, p.185) who propose that 'approximations of practice', not necessarily within an actual classroom, could be used to support trainees' observation skills; this, in turn, might develop their interpretation and understanding of the complexities of classroom interactions (Star et al., 2011).

It has been suggested, for example by Lesgold et al. (1988), that familiarity with the classroom environment leads to better analytical skills of noticing and ultimately an increased 'professional vision' (Sherin, 2000, p.38). This is significant in a context in which a teacher's ability to notice and interpret relevant classroom events has been suggested as being a key component of effective teaching (Gaudin and Chalies, 2015), and which in turn supports 'generative growth', resulting in increased student learning (Franke and Kazemi, 2001, p.105). Jacobs et al. (2010) posit that expertise in such professional noticing is complex, although with sufficient opportunities and support it could be learned. According to some, one way to address this is through the use of video, particularly video observations of practice (e.g. Arya and Christ, 2013); this, it is argued, can scaffold a trainee's flexible interpretation of meaningful student activities and decision-making (e.g. Stürmer et al., 2013), for example by enhancing selective attention (Brunvand, 2010). However, it is important to reflect on what it is trainee teachers need to notice within the context of a lesson observation; not only are the mechanisms of noticing important, but clarity around what it is valuable to notice. One such focus might be the subject-specific pedagogies within the classroom: PCK (Shulman, 1986).

Pedagogical content knowledge

The Carter Review (Carter, 2015) highlights the importance of developing teachers in ways that enable them to 'deconstruct and articulate their practice' (p.41); in particular, it has been argued that trainee teachers should develop a secure grasp of subject-specific pedagogy (termed PCK by Shulman, 1986). PCK refers to the integration and amalgamation of pedagogy and content – the 'what' and 'how' of teaching. Echoing earlier research by Grossman (1990), Barnett and Hodson (2001, highlight PCK as a significant aspect of ITE programmes and continuing professional development, describing it as a critical component of the knowledge needed to teach. This is likely to be predicated on a range of research which indicates that a teacher's PCK can be a significant predictor of student learning and achievement (e.g. Gess-Newsome, 2013) as it integrates teachers' knowledge about curriculum, students' learning and the educational context (Shulman, 1987), thereby allowing teachers to anticipate student difficulties and adaptively respond when students encounter problems (Keller et al., 2016). While there is significant criticism of PCK, not least that it is difficult to define and is an academic construct rather than anything which exists meaningfully within practice (e.g. Loughran et al., 2008), it is favoured by some because it is subject-specific and, as such, requires good subject knowledge (e.g. Park and Chen, 2012). PCK is bound by the requirements of a specific subject and so relies on subject-specific teaching skills and a continual focus on curricular, as well as pedagogical decision-making, something which has been strongly championed within the geography education context (e.g. Brooks, 2017; Healy et al., 2020).

Some researchers indicate that classroom experience is a necessary pre-requisite for the development of PCK (e.g. Cochran et al., 1993; Adams and Krockover, 1997). Indeed, this is suggested by Shulman himself who argues that observations illustrate PCK as the 'management of ideas within a classroom' (Shulman, 1987, p.1), or practice in action, which upholds PCK as having a context-specific nature. Many criticisms of ITE conclude that it does not adequately prepare trainees to link content knowledge and pedagogy to their application in the classroom environment (Blomberg et al., 2011). Further, within a complex ITE landscape, which includes school-centred ITE programmes, there is a growing concern that trainee teachers are not being given sufficient subject-focus within their training year (Tapsfield, 2016). There is a suggestion, therefore, that practices supporting the development of PCK as a mechanism for developing trainee teachers' subject-specific pedagogical understanding should be explored.

Video and immersion

Video viewing is frequently used to help prepare trainee teachers and support their reflective practice (e.g. Wang, 2013; Gaudin and Chalies, 2015). One of the advantages of this approach is that it can provide standardised examples of teaching practice which can be focused to specific areas of learning, without forgoing authenticity (Roche and Gal-Petitfaux, 2015). This helps to link theoretical learning to classroom practice by providing 'springboards for analysis and discussion about teaching and learning' (Borko et al., 2011, p.184). However, standard (sometimes termed 'flat') video has limitations as it is filmed from one angle, often resulting in trainees struggling to fully interpret events (Sherin and Han, 2004).

More recently, immersive 360-degree video has been proposed as having the potential to support trainee teachers to view and identify various features of classroom practices within the atmosphere of being 'really present' (Ibrahim-Didi, 2015; Walshe and Driver, 2018). This presence, which is the main psychological component of immersive virtual environments (Dengel and Mägdefrau, 2018), can elicit more emotional responses, therefore potentially facilitating more memorable learning (Sutcliffe et al., 2005). Within the context of ITE, Walshe and Driver (2018) found the immersive, embodied experience of watching 360-degree video footage supports trainee teachers as it becomes a proxy for real-life classroom settings through which they are able to experience a classroom event, emplaced within its space and time, *being there* in an embodied sense (after Heidegger, 1962) and with agency to select where and with what to engage. They suggest that this 'situatedness' (Gibson, 1977) supports trainees to produce reflections on lessons which show a much better understanding of teacher and pupil behaviours; this then primes them on ways in which they can respond to the immediacy of the context (as suggested by Ibrahim-Didi, 2015; and Craig et al., 2018) and develops in them a better appreciation of pupil engagement and learning. However, Walshe and Driver (2018) identify a need to explore mechanisms by which 360-degree video might be used most effectively with trainee teachers, for example through scaffolded observation opportunities.

It seems then that research stresses the importance of supporting trainee teachers' engagement with and understanding of PCK through lesson observation, but at the same time it highlights a deficit of scaffolded approaches for facilitating meaningful practice in lesson observation per se. While recent work has identified the importance of subject-specificity in lesson observations by mentors (Puttick, 2019; Healy et al., 2020), there appears to be a scarcity of research around trainee teachers' engagement with subject (as demonstrated through PCK) when they are observing lessons. Although flat video has been used as a proxy for classroom observation for some time, these standard approaches may not provide the 'spatial and temporal situatedness required to help [student] teachers to draw on their body-based reflective capabilities in the moment' (Ibrahim-Didi, 2015, p.240). The broader aims of this research are, therefore, to address these gaps by considering how 360-degree video can be used to support trainee teachers to develop their PCK, particularly through creation of what we term 360-degree immersive experiences. This chapter reports on preliminary research findings which address the research question: how do 360-degree immersive experiences support trainee teachers' lesson observation, with a view to developing their PCK?

Project design

This broader research project was framed as an interpretive case study undertaken with 23 final year students on a BA Primary Education Studies course; we adopted Stake's (1994) instrumental case study approach using examination of a particular context to facilitate wider understanding. In this particular instance, the work was aligned with modules developing students' geography, religious education, English and maths PCK across Key Stages 1 and 2 (children of ages 5–7 and 7–11 respectively). It comprised four stages:

Stage 1: Teaching recorded with 360-degree video

We recorded six practising teachers from across two primary academies; sampling was purposive as we recorded in schools in which we had prior relationships through wider partnership activities. Teaching comprised 30-minute lessons of either geography, religious education, English or maths and was recorded using two cameras: a high-end stereoscopic 360-degree camera with omnidirectional audio which was mounted on a monopod and placed at pupils' eye level in the centre of the classroom, and a consumer-level 360-degree monoscopic camera suspended from the classroom's overhead projector. Stereoscopic video provides slightly different images to each eye, adding a natural sense of depth that mimics the way we see. Omnidirectional recording is a method of recording sound that uses multiple microphones (in this case, four), arranged with the purpose of creating 3D stereo sound; this means that when 'watching back' the footage using VR and a headset, attention can be drawn by sound, as well as movement, which gives a more embodied sense of being in the classroom.

Stage 2: Post-teaching teacher reflection

Two weeks after their teaching, teachers reflected on their practice whilst watching the 360-degree video footage. Reflection was unstructured and used a 'think-aloud protocol' which involves participants saying whatever comes into their mind as they complete a task, with the aim of giving observers insight into their cognitive thought processes. Through this, teachers were asked to observe their teaching, note their pedagogical decisions and articulate their thoughts and feelings to facilitate reflection (Cotton and Gresty, 2006). Teacher reflections were recorded using standard video with audio recording for backup.

Stage 3: Creation of interactive 360-degree experiences in virtual reality

The third stage involved the creation of interactive 360-degree experiences in virtual reality (VR) using an immersive overlay platform (IOP). Erikson (2007) identifies a need for careful planning using video as a result of the perceived overwhelming nature of video watching and to support trainees' development of noticing. Various suggestions have been made to address this, for example using short clips with embedded instruction (Brunvand, 2010), or providing scaffolding to identify salient points which 'inspire habits of praxis' (Hewitt et al., 2003, p.500). IOPs facilitate these approaches, allowing identification of important moments within a lesson to be highlighted to draw a viewer's attention through the creation of a navigable interactive multimedia environment. In this case, we incorporated analytical commentaries from the teachers (as standard audio and video clips), posed questions at different points within the teaching, and provided interactive activities that support students to consider teaching and learning from different perspectives (for example engaging with a segment of the lesson plan). These 360-degree experiences were then integrated into the course virtual learning environment; students were able to watch these on desktop computers at home, but more effectively through VR headsets at university, which enabled them to view the media in a more immersive form.

Stage 4: Individual student interviews

During the final stage of the work, we undertook individual, semi-structured interviews with a sample of four trainee teachers to explore their experience with using the 360-degree video and its impact on their understanding of practice. While this was only a small number of trainees, this allowed us to explore in depth their perspectives on the process, thereby developing a rich understanding of their experience to inform future practice and research. The audio-recorded interviews were transcribed; transcriptions and open-ended questionnaire responses were submitted to thematic analysis using a process of naturalistic coding to back up our impressions from the interviews, attempting to balance breadth and depth of focus (after Dey, 1993). Through this process a set of classification categories emerging from the data (inductive content analysis) was developed; this was an iterative process undertaken a number of times to increase the validity of the coding.

We followed the British Educational Research Association (BERA) Ethical Guidelines (2018) and obtained ethical approval from our university research ethics panel prior to the commencement of the project. Given the immersive nature of the recording, this necessitated gaining full written consent from parents/guardians and verbal assent from children, in addition to the standard consent process from school teachers and university-based trainee teachers.

Results: Noticing for understanding

This section firstly considers the themes emerging from the trainee teacher interviews with regards to how use of the 360-degree immersive experiences supported their lesson observations; it then moves on to explore implications for practice, taking a thematic approach to exploring how this approach has the potential to support the development of trainee teachers' PCK.

How do 360-degree immersive experiences support trainee teachers' lesson observation?

Results of this project suggest that the immersive, embodied nature of the 360-degree experiences supports trainee teachers' engagement with lesson observations in a number of ways. Perhaps the clearest theme emerging is that of spatial situatedness, as students feel as though they are physically in the classroom when engaging with the 360-degree experiences: the embodied feeling of 'being there' (Heidegger, 1962). For example, one student commented: 'it was really enjoyable being able to sit in on a class…it really did make you feel like you were there'. It appeared that the 360-degree video became a proxy for real-life classroom settings, such that students were able to engage with the lesson in an embodied way but without disturbing the children or teacher, thereby experiencing a more 'realistic' classroom dynamic. Further, the agency trainees have to explore particular spaces within that classroom through the 360-degree experience is significant; one student commented: 'If something else was going on in a different area you could watch that back, whereas if it was just 2D [video] then you'd have to focus on just one part of the room'. This is particularly significant for the context of undergraduate level ITE programmes where it is not always easy to facilitate students' access to classroom environments, particularly in the earliest stages of a trainee's development.

A mechanism through which the immersive, embodied nature of the 360-degree experiences supported trainees' lesson observation was that of supporting students' understandings of the wider narrative of scene-setting within a lesson. For example, one recorded lesson involved the use of a bakery setting for a Year 2 lesson on fractions. As students experienced the classroom, emplaced within its space and time, they were better able to understand the significance of this narrative as a pedagogical decision; for example, one reflected: 'Getting them to think that they're actually in a bakery … setting the scene [was] good because it enticed the children into what they were going to be doing … it was more creative and exciting'. In this way, trainee teachers were able to see how these culturally significant contexts, such as a bakery, could become contexts for mathematical learning, as proposed by van Oers (2010).

Beyond the narrative, the use of the 360-degree experiences allowed trainees to better understand the pedagogical decision-making processes of the teacher, both in their planning and their *in situ* responses to the fluid happenings of the classroom. For example, one student reflected:

> [the commentary] was really helpful. Like when she said that in order to help the child she had to [model] what she wanted them to do, I wouldn't have picked up on that because I was too busy focusing on other things.

The commentary alongside the 360-degree video was particularly useful for supporting trainees to understand the nuances of classroom relationships, for example effective use of a teaching assistant (TA). One trainee observed: 'I didn't see the TA at first ... she was seamless ... aiding the teacher in ways that we don't think about ... she just must be that well integrated with the children'. The addition of commentary within the 360-degree experiences in this way appears to make explicit teachers' subject expertise, in doing so rendering visible their decision-making processes which are, in turn, driven by what Brooks (2017) terms their professional compass.

Implications for practice: A pathway for developing PCK

While these are only the preliminary results of a small project, there are a number of implications for ITE practice emerging. Firstly, through engaging with 360-degree immersive experiences, trainees were beginning to develop a more nuanced understanding of teachers' pedagogical decision-making, both planned and *in situ*. This, in turn, helped them shift their focus in lesson observations from teaching and managing to children and learning, resulting in a transformative process through which they better understood the subject being taught, and how to transform the subject into forms suitable for teaching (as suggested by Martin, 2005). We suggest, therefore, that if there were a range of 360-degree experiences embedded across the ITE curriculum they would support the development of students' understanding of PCK. Secondly, the process of creating the 360-degree experiences itself is of benefit to the practising teachers who are recorded to provide the content. Previous work found that use of 360-degree video supported students' reflection of their practice and, as a result, they developed a more nuanced understanding of their teaching and improved their self-efficacy (Walshe and Driver, 2018). In a similar way, teachers involved in this research were able to reflect more deeply on their practice; for example, one commented 'I really need to think about how I can stop the children being so passive during the modeled write'; and another: 'that group don't work as well together as I thought. They're just being dominated by one child. I need to rethink my seating plan'. As such, this reinforced the value of the use of 360-degree video for developing in-service teachers as 'reflective practitioners', thereby providing important opportunities for them to sustain their professional learning throughout their careers (Brooks, 2017; Lofthouse, 2018).

However, this preliminary study also revealed a number of concerns. Firstly, the extent to which detailed teacher narrative should be incorporated within the 360-degree experiences. While the detailed reflections were of benefit for those students with little practical classroom experience, this level of scaffolding may have been less useful for others. One student suggested:

> I would have liked to watch it without the commentary, as well, because I think that sometimes by indicating the things that you should be looking at you might miss out on things that the teacher didn't see as important but someone coming from an outside perspective might think is.

This highlights a potential disconnect between what an experienced teacher might focus their reflection on, as compared with what a trainee teacher might benefit from hearing about. For example, an experienced teacher might focus on the minutiae of their decision-making in terms of questioning particular students in their class, but a trainee teacher might benefit from understanding how the teacher is effectively drawing upon the links between the geography and science curriculum in their questioning around the water cycle. Further, experienced teachers may draw on tacit knowledge (Dudley, 2013) when reflecting (e.g. knowing how to shape their explanation based on their experience of misconceptions that have arisen in the past). This may not necessarily be helpful for a novice teacher as they are not able to draw on that form of tacit knowledge, but instead will need to use other sources to develop their curricular and pedagogical decision-making. As such, perhaps it is important to consider how scaffolding might be removed (or added) to facilitate progression of individual trainees or differentiate their experiences according to prior experience and need.

The second point relates to the use of two 360-degree video cameras and how these support (or otherwise) the embodiment of students engaging with the 360-degree experiences. The 360-degree experiences included video from both the stereoscopic camera at pupils' eye level and the second camera suspended from the overhead projector, the intention being to allow engagement with the space from different perspectives. However, this created a proprioceptive disjoint, a breakdown in immersion when moving from one viewpoint to another, particularly from the eye-level perspective to the suspended viewpoint. One student commented 'when you're above that is really strange and I couldn't get used to it; I wanted to see everyone, but I couldn't'. As such, consideration is needed as to how to use multiple 360-degree perspectives to provide a continuous embodied classroom experience. Through this project, it became apparent that 360-degree experiences have significant potential not just within ITE, but beyond for teacher development; the next phase of our research aims to develop this in a wider range of subject and school contexts.

One final issue is a logistical concern relating to schools' willingness to engage with the 360-degree recording. There is increasing unease among schools around GDPR and issues of data privacy; this is compounded by an accountability and risk-averse culture in which schools or multi-academy trusts (MATs) are aware of reputation and the self-imposed need to maintain a perfect external image. We approached numerous schools, the majority of which we had existing positive relationships with through other wider partnership work. However, there was very often a reluctance to allow filming in the school, both as a result of concerns around GDPR and, we believe, the unspoken concern that the quality of their teaching and learning would be scrutinised through the process. This raises an interesting question around use of technology within schools; while technology is advancing to allow ever more sophisticated mechanisms with which to support teaching and learning, school systems and stakeholders may not always be open to embracing them.

Conclusion

Lambert (2018) suggests that a central consideration for teachers should be 'How do we best teach this subject?' (p.367); as such, trainee teachers should be supported to understand pedagogical decision-making and how this is inextricably linked to curricular purpose within any given lesson. Hammond et al. (2019) warn that if ITE mentoring is too focussed upon the technical aspects of the classroom, then it is likely that the curricular and pedagogical questions that bring together concern for what is taught and how it is taught in geography classrooms will be neglected. Addressing this area of concern is particularly important in the context of primary foundation subjects, such as geography, where there has been a reduction in teaching time in ITE (Catling, 2017), and yet the quality of geography education within primary classrooms is under greater scrutiny through Ofsted's[1] new Education Inspection Framework (Kinder and Owens, 2019). Based on the preliminary results of the study reported upon here, we suggest that use of 360-degree experiences supports trainees' engagement with and understanding of lesson observations they undertake; this, in turn, has the potential to illustrate the symbiotic link between theory and practice and, in doing so, support the development of trainee teachers' PCK. What is more, extending this research to secondary classrooms might also address the research gap highlighted by Roberts (2000) on 'the processes that bring together the teachers' subject knowledge, professional knowledge, and on the way teachers and students interact in geography' (Brooks, 2018, p.7).

Note

1 Ofsted (the Office for Standards in Education, Children's Services and Skills) is a non-ministerial department of the UK government, reporting to Parliament, and responsible for inspecting a range of educational institutions, including schools, in England.

References

Abell, S.K. and Cennamo, K.S., 2004. Video cases in elementary science teacher preparation. In J. Brophy (ed.), *Using Video in Teacher Education*. Oxford: Elsevier, pp.103–130.

Adams, P.E. and Krockover, G.H., 1997. Beginning science teacher cognition and its origin in the preservice secondary science teacher program. *Journal of Research in Science Teaching*, 34 (6), pp.633–653.

Arya, P. and Christ, T., 2013. An exploration of how professors' facilitation is related to literacy teachers' meaning construction process during video-case discussions. *Journal of Reading Education*, 39 (1), pp.15–22.

Barnett, J. and Hodson, D., 2001. Pedagogical context knowledge: Toward a fuller understanding of what good science teachers know. *Science Education*, 85 (4), 426–453.

Blomberg, G., Stürmer, K. and Seidel, T., 2011. How pre-service teachers observe teaching on video: Effects of viewers' teaching subjects and the subject of the video. *Teaching and Teacher Education*, 27 (7), pp.1131–1140.

Borko, H., Koellner, K., Jacobs, J. and Seago, N., 2011. Using video representations of teaching in practice-based professional development programs, ZDM. *Mathematics Education*, 42 (1), pp.175–187.

British Educational Research Association, 2018. *Ethical Guidelines for Educational Research*. London: BERA.

Brooks, C., 2017. Pedagogy and identity in initial teacher education: Developing a professional compass. *Geography*, 102 (1), pp.44–50.

Brooks, C., 2018. Insights on the field of geography education from a review of master's level practitioner research. *International Research in Geographical and Environmental Education*, 27 (1), pp.5–23.

Brophy, J., 2004. *Using Video in Teacher Education*. Amsterdam: Elsevier.

Brunvand, S., 2010. Best practices for producing video content for teacher education. *Contemporary Issues in Technology and Teacher Education*, 10 (2), pp.247–256.

Calandra, B. and Brantley-Dias, L., 2006. Using digital video for professional development in urban schools: A preservice teacher's experience with reflection. *Journal of Computing in Teacher Education*, 22 (4), pp.137–145.

Carter, A., 2015. *The Carter Review of Initial Teacher Training (ITT)*. London: Department for Education.

Catling, S., 2017. Not nearly enough geography! University provision for England's pre-service primary teachers. *Journal of Geography in Higher Education*, 41 (3), pp.434–458.

Cochran, K.F., DeRuiter, J.A. and King, R.A., 1993. Pedagogical content knowledge: An integrative model for teacher preparation. *Journal of Teacher Education*, 44 (4), pp.263–272.

Cotton, D. and Gresty, K., 2006. Reflecting on the think-aloud method for evaluating e-learning. *British Journal of Educational Technology*, 37 (1), pp.45–54.

Craig, C.J., You, J., Zou, Y., Verma, R., Stokes, D., Evans, P. and Curtis, G., 2018. The embodied nature of narrative knowledge: A cross-study analysis of embodied knowledge in teaching, learning, and life. *Teaching and Teacher Education*, 71, pp.329–340.

Dengel, A. and Mägdefrau, J., 2018. Immersive learning explored: Subjective and objective factors influencing learning outcomes in immersive educational virtual environments. In *2018 IEEF International Conference on Teaching, Assessment and Learning for Engineering (TALE): Woollongong, Australia, 4–7 December 2008*. Available at: https://ieeexplore.ieee.org/xpl/conhome/8600698/proceeding [Accessed 19 December 2019].

Dey, I., 1993. *Qualitative Data Analysis: A User-Friendly Guide for Social Scientists*. Abingdon: Routledge.

Dudley, P. 2013. Teacher learning in lesson study: What interaction-level discourse analysis revealed about how teachers utilised imagination, tacit knowledge of teaching and fresh evidence of pupils learning, to develop practice knowledge and so enhance their pupils' learning. *Teaching and Teacher Education*, 34, pp.107–121.

Erikson, F., 2007. Ways of seeing video: Toward a phenomenology of viewing minimally edited footage. In R. Goldman, R. Pea, B. Baron and S. Derry (eds), *Video Research in the Learning Sciences*. Mahwah, NJ: Erlbaum, pp.145–155.

Franke, M.L. and Kazemi, E., 2001. Learning to teach mathematics: Focus on student thinking. *Theory into Practice*, 40 (2), pp.102–109.

Gaudin, C. and Chalies, S., 2015. Video viewing in teacher education and professional development. *Educational Research Review*, 16, pp.41–67.

Gess-Newsome, J., 2013. Pedagogical content knowledge. In J. Hattie and E.M. Anderman (eds), *International Guide to Student Achievement*. New York: Routledge, pp. 257–259.

Gibson, J.J., 1977. The theory of affordances. In R. Shaw and J. Bransford (eds), *Perceiving, Acting, and Knowing*. Mahwah, NJ: Lawrence Erlbaum Associates.

Grossman, P., 1990. *The Making of a Teacher: Teacher Knowledge and Teacher Education*. New York: Teachers College Press.

Grossman, P. and McDonald, M., 2008. Back to the future: Directions for research in teaching and teacher education. *American Educational Research Journal*, 45 (1), pp.184–205.

Hammond, L., Mitchell, D. and Palomba, M., 2019. Mentors in geography education: An under-used resource and under-represented community? *Teaching Geography*, 44 (1), p.6.

Healy, G., Walshe, N. and Dunphy, A., 2020. How is geography rendered visible as an object of concern in written lesson observation feedback? *The Curriculum Journal*, 31 (1), pp.7–26.

Heidegger, M., 1962. *Being and Time*. New York: Harper and Row.

Hewitt, J., Pedretti, E., Bencze, L., Vaillancourt, B.D. and Yoon, S., 2003. New applications for multimedia cases: Promoting reflective practice in preservice teacher education. *Journal of Technology and Teacher Education*, 1 (4), pp.483–500.

Ibrahim-Didi, K., 2015. Immersion with 360 video settings: Capitalising on embodied perspectives to develop reflection-in-action within pre-service teacher education. In T. Thomas, E. Levin, P. Dawson, K. Fraser and R. Hadgraft (eds), *Research and Development in Higher Education: Learning for Life and Work in a Complex World* (Vol. 38). Melbourne: HERDSA, pp.235–245.

Jacobs, V., Lamb, L. and Phillip, R., 2010. Professional noticing of children's mathematical thinking. *Journal for Research in Mathematics Education*, 41 (2), pp.169–202.

Keller, M.M., Neumann, K. and Fisher, H.E., 2016. The impact of physics teachers' pedagogical content knowledge and motivation on students' achievement and interest. *Journal of Research in Science Teaching*, 54 (5), pp. 586–614.

Kinder, A. and Owens, P., 2019. The new Education Inspection Framework through a geographical lens. *Primary Geography*, 100, pp.10–13.

Lambert, D., 2018. Teaching as a research-engaged profession: Uncovering a blind spot and revealing new possibilities. *London Review of Education*, 16 (3), pp.357–370.

Lesgold, A., Rubinson, H., Feltovitch, P., Glaser, R., Klopfer, D. and Wang, Y., 1988. Expertise in a complex skill: Diagnosing x-ray pictures. In M.T.H. Chi, R. Glaser and M. Farr (eds), *The Nature of Expertise*. Hillsdale, NJ: Erlbaum, pp.311–342.

Lofthouse, R.M., 2018. Re-imagining mentoring as a dynamic hub in the transformation of initial teacher education: The role of mentors and teacher educators. *International Journal of Mentoring and Coaching in Education*, 7 (3), pp.248–260.

Loughran, J., Mulhall, P. and Berry, A., 2008. Exploring pedagogical content knowledge in science teacher training. *International Journal of Science Education*, 30 (10), pp.1301–1320.

Luna, M.J. and Sherin, M.G., 2017. Using a video club design to promote teacher attention to students' ideas in science. *Teaching and Teacher Education*, 66, pp.282–294.

Martin, F., 2005. The Relationship between Beginning Teachers' Prior Conceptions of Geography, Knowledge and Pedagogy and their Development as Teachers of Primary Geography. PhD dissertation, University of Coventry.

Mattsson, M., Eilertson, T. and Rorrison, D., 2011. *A Practicum Turn in Teacher Education*. Rotterdam: Sense Publishers.

Murray, J. and Mutton, T., 2016. Teacher education in England: Change in abundance, continuities in question. In Teacher Education Group (ed.), *Teacher Education in Times of Change*. Bristol: Policy Press, pp.57–74.

Park, S. and Chen, Y.-C., 2012. Mapping out the integration of the components of pedagogical content knowledge (PCK): Examples from high school biology classrooms. *Journal of Research in Science Teaching*, 49 (7), pp.922–941.

Puttick, S., 2019. Geography teachers' written lesson observation feedback. Presentation given at Geography Teacher Educators' Conference, Bristol, UK, 24–26 January.

Roberts, M., 2000. The role of research in supporting teaching and learning. In W.A. Kent (ed.), *Reflective Practice in Geography Teaching*. London: Sage, pp.287–296.

Roche, L. and Gal-Petitfaux, N., 2015. A video-enhanced teacher learning environment based on multimodal resources: A case study in PE. *Journal of E-Learning and Knowledge*, 11 (2), pp.91–110.

Scherr, R.E. and Hammer, D., 2009. Student behaviour in epistemological framing: Examples from collaborative active learning activities in physics. *Cognition and Instruction*, 27 (2), pp.147–174.

Sherin, M.G., 2000. Viewing teaching on video tape. *Educational Leadership*, 57 (8), pp.36–38.

Sherin, M.G. and Han, S.Y., 2004. Teacher learning in the context of video club. *Teaching and Teacher Education*, 20 (2), pp.163–183.

Sherin, M.G. and van Es, E.A., 2005. Using video to support teacher's ability to interpret classroom interactions. *Journal of Technology and Teacher Education*, 13 (3), pp.475–491.

Shulman, L.S., 1986. Those who understand: Knowledge growth in teaching. *Educational Researcher*, 15 (2), pp.4–14.

Shulman, L.S., 1987. Knowledge and teaching: Foundations of the new reform. *Harvard Educational Review*, 57 (1), pp.1–22.

Stake, R.E., 1994. Case studies. In N.K. Denzin and Y.S. Lincoln (ed.), *Handbook of Qualitative Research*. Thousand Oaks, CA: Sage, pp.236–247.

Star, J.R., Lynch, K. and Perova, N., 2011. Using video to improve mathematics teachers' abilities to attend to classroom features: A replication study. In G. Miriam, V. Sherin, R. Jacobs and R.A. Phillip (eds), *Mathematics Teachers' Noticing: Seeing through Teachers' Eyes*. New York: Routledge, pp.117–133.

Stürmer, K., Konings, K.D., Seidel, T. and Schäfer, S., 2013. Declarative knowledge and professional vision in teacher education: Effect of courses in teaching and learning. *British Journal of Educational Psychology*, 83 (3), pp.467–483.

Sutcliffe, A., Gault, B. and Shin, J.-E., 2005. Presence, memory and interaction in virtual environments. *International Journal of Human Computer Studies*, 62 (3), pp.307–327.

Tapsfield, A., 2016. Teacher education and the supply of geography teachers in England. *Geography*, 101 (2), pp.105–109.

Tripp, T.R. and Rich, P.J., 2012. The influence of video analysis on the process of teacher change. *Teaching and Teacher Education*, 28(5), pp.728–739.

van Oers, B., 2010. Emergent mathematical thinking in the context of play. *Educational Studies in Mathematics*, 74 (1), pp.23–37.

Walshe, N. and Driver, P., 2018. Developing reflective trainee teacher practice with 360-degree video. *Teaching and Teacher Education*, 78, pp.97–105.

Wang, X., 2013. A potential approach to support pre-service teachers' professional learning: The video analysis of the authentic classroom. *US–China Education Review B*, 3, pp.149–161.

Chapter 5

Children, childhood and children's geographies
Evolving through technology

Lauren Hammond

Introduction

Children are central to education, and education is often a central part of children's lives. On initial reading this statement seems unproblematic. However, the relationships between children's everyday lives, geographies and knowledge, and the specialist knowledge they engage with in schools as part of their formal education, is much debated. These debates are philosophical, for example considering the purpose of schooling (Young et al., 2014) and its potential for emancipation (Freire, 1970) and human flourishing (Reiss and White, 2013). They are also highly practical, as teachers engage with what Lambert and Morgan (2010) term 'curriculum making' – which represents the process in which 'the curriculum comes in to being via the day-to-day interactions between teachers, their students and the subject discipline' (Lambert and Biddulph, 2015, p.215).

Geography is in a unique position in relation to these debates, as a major area of research in the academic discipline is everyday life. This includes the study of children's and young people's geographies. Tani (2011) argues the importance of these debates for geography as a school subject, and asserts that geography is one of the few spaces in the school curriculum 'in which students' experiences and relationships with their environments can be taken into account' (p.27). However, despite Tani and others (Young People's Geographies Project, 2011; Biddulph, 2012; Yarwood and Tyrell, 2012; Catling, 2014; Roberts, 2017) extolling the benefits of engaging with ideas and methodologies from the academic discipline to actively consider children's geographies in the school classroom, barriers often exist which prevent this from happening in practice (Catling, 2011). These barriers can be multi-faceted and commonly include: time and space in the curriculum, teacher education and teachers' knowledge of children's geographies, and the existence of accountability and performativity pressures in schools (Catling, 2011; Hammond, 2020).

The context of a digital world (Walshe and Healy, 2020) brings new dimensions and areas of consideration to these discussions. This is because technology is changing children's lives and geographies, as well as perceptions and representations of childhood. Examining these debates is of value to geography education in developing teachers' knowledge of the children they teach and considering the social contexts that both they, and their students, exist within and contribute to. This knowledge can both inform and support teachers as they engage in curriculum making.

This chapter considers these debates, specifically focusing on the development of Web 2.0 (including social media) since the 1990s. It argues that if geography education in schools fails to critically consider and engage with children's geographies (including their experiences of a digital world), then it risks creating what Freire (1970) terms 'banking education'. In this situation 'education becomes an act of depositing, in which the students are the depositories and the teacher is the depositor' (Freire, 1970, p.45). Students are conceptualised only as being able to receive information that the teacher provides. Thus, children's opportunities for meaning-making and engaging in student–teacher reciprocal dialogue is limited, along with respect for, and engagement with, children's everyday knowledge and geographies.

In considering children's rich and varied experiences and imaginations of the world, this chapter begins by examining what is meant by childhood and children's geographies. It then examines how developments in digital technology over the last 30 years have changed both children's lives and the 'production of space' (Lefebvre, 1991). Following this, the chapter draws upon and shares the narratives of young people who participated in research I conducted as part of my doctorate. This section focuses specifically on sharing children's experiences and perceptions of a digital world. The chapter then concludes by considering how and why these debates are important for school geography, and raises questions for consideration by geography educators to move these debates forward.

What is childhood, and what are children's geographies?

The concepts of children and childhood are familiar to most people, and are often ingrained in shared social imaginations of the world. Indeed, they can be so familiar to us that they can seem 'natural' (Matthews and Limb, 1999; Aitken, 2001; Skelton, 2008). However, 'childhood is a contested notion' (Freeman and Tranter, 2015, p.491) and children are not a homogenous group. Debates in the academy now acknowledge children and childhood as being socially constructed and historically situated (rather than solely biologically defined; Valentine et al., 1998). Furthermore, children are recognised as having an active role in constructing their own social identities (Skelton, 2008) and as contributing to the production of social space (Lefebvre, 1991; Hammond, 2019).

Research into and debates about children and young people were absent for much of the early development of geography as an academic discipline (Aitken, 2001; Freeman and Tranter, 2015). In 1970s North America, academics including Bunge and Bordessa began to examine the everyday geographies and spatial repression of children (Aitken, 2001). This work was both fuelled by and informed wider socio-political debates, for example about inequality, as well as discourse in the academy about the role of geography in researching and representing all people(s) (Peet, 2015).

From its emergence as a sub discipline, a key area of concern in children's geographies has been not only to further knowledge about children's experiences and imaginations of the world, but to provide opportunities for children to share their voices. This philosophy has informed the design of participatory and emancipatory methodologies, which have been developed in the field (van Blerk et al., 2009); these critically consider the ethics, politics and power relations of working with children and young people (Valentine et al., 1998). These philosophies and debates also inform discourse as to whether – and how – young people are able to participate in their communities (McKendrick, 2009).

Today, children's geographies is a vibrant and growing sub discipline. Research into and debates about children's geographies are diverse, international and interdisciplinary, with other socio-political groups such as policy makers and non-governmental organisations also engaging with debates in the field (Holloway and Pilmott-Wilson, 2011; Holt, 2011).

However, despite advances in the sub discipline, concerns have also been raised around the extent to which children's geographies sometimes acts as a gated community (Horton et al., 2008; Holt, 2011). This can be seen to have resulted in some knowledge, methodologies and debates about children's geographies remaining within the confines of the sub disciplines' dedicated conferences and journals (Horton et al., 2008; Holt, 2011). If we consider knowledge about children's geographies to be of value to geography teachers in their curriculum making, then this raises important philosophical and practical questions for geography as a school subject. These include critically considering how and why school geography can, and should, access, engage with, and use knowledge about children's geographies.

Before considering how and why ideas might be shared between children's geographies and geography education, the chapter introduces the context of a digital world. It focuses specifically on how changes in technology have changed the lives and geographies of children and young people. This not only draws on what is written in academic literature, but also listens to the perspectives of young people themselves.

Changing technology, changing children's geographies?

Although now thoroughly ingrained in the social fabric and everyday lives of a substantial proportion on the world's population, the World Wide Web (WWW) is a fairly recent phenomenon. It was invented in 1991, with Web 2.0 emerging early in the new millennium. The advent of Web 2.0 has transformed how information is shared, disseminated, consumed and is responded to (Kitchin et al., 2013); it has shifted communication to be increasingly interactive and two-way, and has influenced and changed interaction at individual, community and societal levels (Van Dijck, 2013). Put another way, these shifts in technology have resulted in significant changes to people's everyday lives, and how they choose to represent and share their own lives, as well as how they are represented by others. These changes have also altered social space and how it is produced, as well as conceptions and representations of time.

Considerable socio-political debate about the impacts of living in a digital age has ensued since the advent and growth of Web 2.0, including social media. Livingstone and Bovill (2002) conceptualise the debates as between optimists and pessimists. They argue optimists perceive that technological advancements offer new opportunities for creativity and play, and also opportunities to enhance democracy and to provide a voice to people who have previously been under-represented (e.g. children and young people). In contrast, pessimists raise concerns about the potential for these technological changes to lead to the challenging of authority and traditional values, as well arguing that they may lead to social impacts from a different, and perhaps more sedentary, lifestyle (Livingstone and Bovill, 2002).

In line with the widely argued social construction of children as angels (who are vulnerable and need protecting) and devils (who need controlling) (Valentine, 1996), debates about the impacts of technology have often had a 'pessimistic' (Livingstone and Bovill, 2002) bias when they focus on children and young people. In a variety of socio-political spaces, concerns have been raised about technological change and the context of a digital world affecting both children's lives and parenting (Plowman et al., 2010; Livingstone and Smith, 2014; Ofcom, 2017). For example, Livingstone and Smith (2014) highlight some of the concerns of children engaging with social media as being: 'cyberbullying, contact with strangers, sexual messaging ("sexting") and pornography' (p.636). They argue that these concerns have, at times, become the subject of large-scale political and public debates which attract the interest of a variety of people including parents and carers, educationalists and clinicians (Livingstone and Smith, 2014).

Despite research rarely making positive connections between social media use and desirable outcomes for children and young people (Plowman et al., 2010), some benefits of children engaging with a digital world are acknowledged and extolled. These include an increased access to a more diverse range of information, increased opportunities for sharing one's voice, active participation in different social and political forums, and new spaces of play being created. Glascott Burris and Wright (2001) also suggest that changes in technology have changed dynamics between the adult and the child, with children sometimes having more extensive experience and superior knowledge of technology than adults around them.

Although it is recognised that children are not a homogenous group, and therefore that they will engage with technology in a multitude of different ways, the scale of social changes that have been born out of the advent and growth of Web 2.0, including social media, are worthy of examination. In the case of the United Kingdom (UK), Ofcom (2017) report that 99% of young people aged between 12 and 15 (the age range of the young people in the research examined later in this chapter) go online for at least 21 hours per week, 90% use YouTube and 74% have a social media profile. Although this data does not tell us why, or how, young people are engaging with the internet, or the impacts it has, or has had, on their geographies, it helps us to conceptualise Web 2.0 as being ingrained in young people's daily lives in the UK. The social natures of the platforms young people are engaging with also represent changes in both the production of space, and representation of life and being, through technological change.

Thus far in this chapter, I have introduced the concept of the child and the sub discipline of children's geographies. Following this, I examined debates about how children's geographies and conceptions of childhood have changed with, and through, technological advancements. However, until now children's voices have been missing from this chapter. In line with one of the primary aims of the sub discipline of children's geographies – to conduct research with and for children, and to empower and enable children to share their voices (van Blerk et al., 2009) – I now introduce my doctoral research. Following this, I share narratives of young people who took part in the research, about their experiences and perceptions of living in a digital world.

Introducing the research

My doctoral research was an investigation into children's geographies and their value to geography education in schools. Recognising that children have different experiences and imaginations of the world (Matthews and Limb, 1999), which may also be difficult for adults to fully understand, I began my research by listening to children's voices. I did this by collecting data from a series of six semi-structured interviews with a group of five young people, aged 13, in London, through a 'storytelling and geography group'.

The methodology drew on Goodson's (2013) work on life histories, and encouraged young people to share their geographies and imaginations of London. Life histories research involves the triangulation of oral data with the historical context and other narratives. This enables active consideration of how individual (and private) narratives interweave with public (and shared) narratives (Goodson, 2013). The value of this, and the group nature of the interviews, lies in making connections between the geographies of individuals and broader socio-cultural narratives (such as the evolution of a digital world), which the young people both shape and are shaped by (Cameron, 2012).

Following data collection, the recordings of the 'storytelling and geography group' were transcribed. They were then inductively coded to condense the data and identify themes for further examination. Whilst this process enabled me to cluster and categorise data, overlap between themes led to their potential under-examination. This led me to engage in a second cycle of coding, in which I used Harvey's (1990) 'grid of spatial practices' (see Figure 5.1) to code the narratives within and across the themes from the first cycle of coding.

Harvey's (1990) grid of spatial practices was developed from and informed by Henri Lefebvre's (1991) work on the production of space. Motivated by what he described as his 'critical conscience' on everyday life (Elden, 2006, p.190), Lefebvre sought to 'grasp how the production of space, patterns of state spatial organization, and geographies of socio-political struggle are being reshaped under late twentieth-century capitalism' (Brenner and Elden, 2009, p.25). In his book *The Production of Space* (translated into English from his native French in 1991), Lefebvre introduces a conceptual triad to support the consideration of how space is produced, sustained and evolves. The dimensions of this triad are: spatial practices, representations of space and representational space. Harvey's definitions of these terms are included in Figure 5.1. However, to facilitate further examination of the complexities and subtleties of spatial practices in urban settings, Harvey (1990) adds three further dimensions to Lefebvre's triad (Figure 5.1):

> Accessibility and distanciation speak to the role of the 'friction of distance' in human affairs. Distance is both a barrier and a defense against human interaction. It imposes transaction costs upon any system of production and reproduction (particularly those based on any elaborate social division of labor, trade, and social differentiation of reproductive functions). Distanciation is simply a measure of the degree to which the friction of space has to be overcome to accommodate social interaction;
>
> The appropriation of space examines the way in which space is used and occupied by individuals, classes, or other social groupings. Systematized and institutionalized appropriation may entail the production of territorially founded forms of social solidarity;
>
> The domination of space reflects how individuals or powerful groups dominate the organization and production of space so as to exercise a greater degree of control either over the friction of distance or over the manner in which space is appropriated by themselves or others. (Harvey, 1990, pp.258–259)

Harvey explains that the 'dimensions of the grid are not independent of each other' (p.259). For example, any domination of space may lead to some people feeling a friction of distance from others, or the place and/or time-space they exist within. The work of Lefebvre (1991) and Harvey (1990) are of value in analysing this research, as they enable examination of how young people are shaped by – and shape – social space. In considering the context of a digital world, this can provide insight into young people's lives and geographies, as well as enabling us to consider their imaginations of technology and the digital world(s) they exist within and contribute to.

I now move on to examine the narratives of the young people in the research, specifically focusing on those reflecting their experiences and imaginations of a digital world.

	Accessibility and distanciation	Appropriation and use of space	Domination and control of space
Material spatial practices (experience)	Flows of good, money, people, labour power, information etc; transport & communications systems, market and urban hierarchies; agglomeration	Urban built environment, social space and other 'turf' designations; social networks of communication & mutual aid	Private property in land, state, & administrative divisions of space, exclusive communities & neighbourhoods, exclusionary zoning & other forms of social control (policing and surveillance)
Representations of space (perception)	Social, psychological and physical measures of distance, mapmaking; theories of the 'friction of distance' (principle of least effort, social physics, range of good, central place and other forms of location theory)	Personal space; mental space; spatial hierarchies, symbolic representation of spaces	Forbidden space "territorial imperatives", community, regional, culture, nationalism, geopolitics, hierarchies
Spaces of representation (imagination)	"Media is the message" new modes of spatial transaction (radio, TV, film, photography, painting etc); diffusion of "taste"	Popular spectacles– street demonstrations, riots; places of popular spectacle (streets, squares, markets); iconography and graffiti	Organized spectacles, monumentality and constructed spaces of ritual; symbolic barriers and signals of symbolic capital

Material spatial practices
Refer to the physical and material flows, transfers and inter-actions that occur in and across space in such a way as to assure production and social reproduction

Representations of space
Encompass all of the signs and signifiers, codes and knowledge, that all such material practices to be talked about and understood, no matter whether in terms of everyday common sense or through the arcane jargon of the academic disciplines that deal with spatial practices (engineering, architecture, geography, planning, social ecology, and the like)

Spaces of representation
Are social inventions (codes, signs, and even material constructs such as symbolic spaces, particular built environments, paintings, museums and the like) that seek to generate new meanings and possibilities for spatial practices

Figure 5.1 Harvey's grid of spatial practices.
Source: Harvey (1990, p.257).

Children's narratives about living in a digital world

A key finding of the research was that the children in the study navigated multiple, sometimes contradictory, social spaces when constructing and representing themselves and their identities in London. The context of a digital world featured strongly in two themes related to identity in the analysis. These were:

Voice and identity: This theme included narratives related to the use of social media to share experiences of being and life.

Gender, sex, sexuality and identity: This theme included narratives in which the children considered access to what Harvey's (1990) grid terms 'forbidden spaces' (e.g. pornography and information about crimes) via the internet. Narratives also included consideration of how sex, sexuality and gender are represented on and through (social) media (including music videos), and also using social media to share and learn about others' relationship status.

The remainder of this section will focus on sharing the narratives of young people who took part in the research on the two themes highlighted above, beginning with voice and identity. All names given are pseudonyms. Following this, I will move on to consider the value of these debates to school geography.

One area of discussion that emerged during the 'storytelling and geography group' concerned how and why young people share their lives and stories with others. In the narrative below, Jessica talks about a gang in her locality, who she explains are 'reppin' for' (representing) their lives and area through a rap they have uploaded to YouTube:

JESSICA: While that one, it doesn't really, but it does set a bad example for London in a way. Because people are just gonna think, it's just full of teenagers, with alcohol, and rappers, and stuff like that. But most places where you go, will have like a gang, that rap about their area, you know what I'm sayin'?
RESEARCHER: Why do you think that is?
JESSICA: Because like, they want people to know, like people that aint like us, like the Prime Minister or something, but they aint gonna listen to it are they? But basically, what I'm trying to say, this is my opinion, I think they're trying to let people know about our area, and how they grow up and stuff.
RESEARCHER: And what do they say that their lives are like?
JESSICA: Some of them say that their life's been hard, and also how they got into the gang, and they rap about what they do in the gang and stuff.

Jessica's narrative can be interpreted as her expressing a perspective that young people are often not listened to by powerful people and institutions within society. Her narratives can be seen to reflect a perspective that the gang feel and are expressing a 'friction of distance' (Harvey, 1990; Figure 5.1) from society because of who they are and their socio-economic backgrounds. They can be seen to reflect a perceived lack of voice in societal debates, which is, as highlighted earlier in this chapter, a fundamental area of interest for children's geographies. However, the context of a digital world, and in this case YouTube, can be interpreted as providing them with both a platform to – and perhaps a sense of opportunity and hope that they can – share their lives and voices with others, including those who they perceive to be powerful, and with whom they may not have previously communicated.

The second theme analysed as related to identity in a digital world, is gender, sex and sexuality. Like voice, gender and sexuality are often fundamental to a person's identity (Jackson, 1992). However, Brown and Browne (2016) argue that despite the fact they have often been present, they have rarely been explicitly addressed in human geography. Further to this, children's experiences and imaginations of sex and sexuality are often perceived as an 'uncomfortable' topic for many people (Brown and Browne, 2016). For example,

Anglo-European cultures often socially repress discussions about sex and sexuality which include reference to children (Foucault, 1978; Aitken, 2001), with Valentine et al. (1998, p.24) arguing that there has been 'very little consistent research on questions of sex, sexuality and gender' related to young people. Valentine et al. (1998) go on to state that the research that has been undertaken often relates to preventing the spread of sexually transmitted infections (STIs) or stopping teenage pregnancies, as opposed to exploring young people's perceptions and experiences of sex, sexuality and gender.

Throughout the 'storytelling and geography group', the young people shared multiple narratives that related to this theme. A large proportion of these narratives can be interpreted as relating to the context of a digital world. For example, the young people discussed using social media to share their relationship status and accessing pornography websites via their mobile phones. As reflected in the narrative below which focuses on Nicki Minaj's song 'Anaconda', they also consider how gender, sexuality and sex are represented in the media:

JACK: It was just a smoking guy, some old guy. Miss, you see when Jessica was talking about the guy, and people rapping about stuff, and people talking about sex. You see that probably got a million views, 'Anaconda' the new song, she's just showing her cleavage and her arse.
JESSICA: Her arse.
TILLY: She's famous already.
JACK: That got, that got, in two hours that got 300 views. 3 million views sorry.
JESSICA: You don't have to show your arse, and your boobs, and your cleavage and everything, and your belly and your legs.
TILLY: I think it's kind of sexist!
JACK: The video is so, the video is so bad!
JESSICA: You just don't have to show your legs and stuff just to get famous and just to get loads of views on it.
RACHEL: Nowadays the majority of people do.
TILLY: Yeah, you do kind of have to do that!

This discussion can be read as the young people critically considering the portrayal and representations of gender and sex in the song 'Anaconda'. The group discuss complex social issues such as the ethics related to Minaj's singing, dancing and dressing in a sexually provocative manner, considering further as to whether there is a social pressure for women to do this to 'get views' (on media and social media platforms) and to become famous.

These two examples of children's narratives about dimensions of identity in a digital world highlight the complexities of navigating and shaping a variety of social spaces as a young person, in this case in London. I now move on to consider the value of children's geographies to geography education in schools before concluding the chapter.

Towards valuing children's geographies in geography education in schools

Children's geographies and how they change (for example, through a digital world) raise important pedagogical questions for geography education in school; these include how and why teachers consider, value and connect to children's everyday knowledge and geographies in the classroom. Considering these questions is significant, both in supporting students with meaning-making as they connect to specialist geographical knowledge in schools (Roberts, 2017), and also in respecting children's rich and varied experiences and imaginations of the world, and sometimes supporting children in questioning and deconstructing their views.

They also raise important questions about curriculum. As geography as a discipline studies everyday life and children's geographies, why should this not be an area of study in schools? Lambert and Morgan (2010) point out that school geography has at times been 'socially selective' about what is included in the curriculum. Education is always political (Catling, 2014), but it is of value to consider why children's geographies have often been excluded from school geography. Although this may be the result of fields of knowledge sometimes acting as 'gated communities', it is also worthy of critical consideration as to whether, how and why children have, at times, been subordinated by society and schooling (Freire, 1970; Foucault, 1978; Catling, 2014; Giddens, 2016).

I argue that these questions are of value in paving paths to cross 'borders' (Castree et al., 2007) between fields of knowledge (in this case geography education and children's geographies; Hammond, 2020), and also for teachers making decisions about curriculum, pedagogy and purpose as they engage in curriculum making. These debates are pertinent, as although students and their experiences and everyday knowledge are often included and represented in many models and approaches to teaching geography (e.g. Bennetts' (2005) 'roots of understanding' model; Lambert and Morgan's (2010) 'curriculum making' model; the GeoCapabilities (2016) approach), if children's geographies (both as shared by children themselves and the sub discipline) are not considered in school geography, then we risk constructing a banking model of education. This concern is particularly relevant in the context of the accountability and performativity pressures that schools now face (Biddulph et al., 2015), and the questions raised here aim to offer some suggestions for consideration to support these debates in moving forward.

Conclusion

This chapter has highlighted that children's geographies and imaginations of the world are rich and varied, as is the study of them in the sub discipline of children's geographies. The context of a digital world is changing the lives and geographies of children and young people in diverse ways, as well as changing the production of space. It is offering not only opportunities (e.g. new spaces of play, access to information and opportunities for voice), but also challenges and dangers (e.g. sexting, open access to hardcore pornography and a more sedentary lifestyle). These changes are both experienced and imagined by children and young people, and they are researched and represented in the academy.

For geography education in schools, the context of a digital world and changing children's geographies offers new opportunities to critically consider how and why children's geographies are respected and explored in the classroom. Although the digital world and the changes it has brought to children's geographies raise some often difficult questions (as they relate to identity and the relationships between adults and children in schools, and potentially changing curriculum and pedagogical approaches), these questions are important to geography education. This is because knowledge of children's geographies, as drawn from both the academic discipline and shared by children themselves, has the potential to make geography teachers more informed in their curriculum making. Further to this, geography offers one of the few spaces in the curriculum where children and young people can engage with and connect to specialised knowledge which examines everyday geographies, thus having the potential to enable children to situate and explain their own geographies (Hammond, 2020).

References

Aitken, S., 2001. *Geographies of Young People: The Morally Contested Spaces of Identity*. London: Routledge.
Bennetts, T., 2005. The links between understanding, progression and assessment in the secondary geography curriculum. *Geography*, 90 (2), pp.152–170.
Biddulph, M., 2012. Spotlight on: Young people's geographies and the school curriculum. *Geography*, 97 (3), pp.155–162.
Biddulph, M., Lambert, D. and Balderstone, D., 2015. *Learning to Teach Geography in the Secondary School: A Companion to School Experience*. Routledge: London.
Brenner, N. and Elden, S., 2009. Introduction, state, space, world: Lefebvre and the survival of capitalism. In N. Brenner, S. Elden and G. Moore (eds), *State, Space and World: Selected Essays*. Minneapolis, MN: University of Minnesota Press, pp.1–50.
Brown, G. and Browne, K., 2016. *The Routledge Companion to Geographies of Sex and Sexuality*. London: Routledge.
Cameron, E., 2012. New geographies of story and storytelling. *Progress in Human Geography*, 36 (5), pp.573–592.
Castree, N., Fuller, D. and Lambert, D., 2007. Geography without borders. *Transactions of the Institute of British Geographers*, 32 (3), pp.317–335.
Catling, S., 2011. Children's geographies in the primary school. In G. Butt (ed.), *Geography, Education and Future*. London: Bloomsbury, pp.15–28.
Catling, S., 2014. Giving younger children voice in primary geography: Empowering pedagogy: A personal perspective. *International Research in Geographical and Environmental Education*, 23 (4), pp.350–372.
Elden, S., 2006. Some are born posthumously: The French afterlife of Henri Lefebvre. *Historical Materialism*, 14 (4), pp.185–202.
Foucault, M., 1978. *The Will to Knowledge: The History of Sexuality* (Vol. 1). Oxford: Penguin Books.
Freeman, C. and Tranter, P., 2015. Children's geographies. In D.J. Wright (ed.), *International Encyclopedia of the Social and Behavioral Sciences* (2nd edn). London: Elsevier, pp.491–497.
Freire, P., 1970. *Pedagogy of the Oppressed*. Oxford: Penguin Books.
GeoCapabilities, 2016. *GeoCapabilities*. Available at: www.geocapabilities.org/ [Accessed on 2 September 2019].
Giddens, A., 2016. *The Constitution of Society*. Cambridge: Polity Press.
Glascott Burris, K. and Wright, C., 2001. Review of research: Children and technology: Issues, challenges and opportunities. *Childhood Education*, 78 (1), pp.37–41.
Goodson, I., 2013. *Developing Narrative Theory: Life Histories and Personal Representation*. London: Routledge.
Hammond, L., 2019. Utilising the 'production of space' to enhance young people's understanding of the concept of space. *Geography*, 104 (1), pp.28–38.
Hammond, L., 2020. An Investigation into Children's Geographies and Their Value to Geography Education in Schools. PhD thesis, University College London.
Harvey, D., 1990. Flexible accumulation through urbanisation: Reflections on 'post-modernism' in the American city. *Perspecta*, 26, pp.251–272.
Holloway, S. and Pilmott-Wilson, H., 2011. Geographies of children, youth and families: Defining achievements, debating the agenda. In L. Holt (ed.), *Geographies of Children, Youth and Families: An International Perspective*. London: Routledge, pp.9–24.

Holt, L., 2011. Introduction: Geographies of children, youth and families: Disentangling socio-spatial contexts of young people across the globalising world. In L. Holt (ed.), *Geographies of Children, Youth and Families: An International Perspective*. London: Routledge, pp.1–8.

Horton, J., Kraftl, P. and Tucker, G., 2008. The challenges of "children's geographies": A reaffirmation. *Children's Geographies*, 6 (40), pp.335–348.

Jackson, P., 1992. *Maps of Meaning: An Introduction to Cultural Geography*. Abingdon: Routledge.

Kitchin, R., Linehan, D., O'Callaghan, D. and Lawton, P., 2013. Public geographies through social media. *Dialogues in Human Geography*, 3 (1), pp.56–72.

Lambert, D. and Biddulph, M., 2015. The dialogic space offered by curriculum-making in the process of learning to teach, and the creation of a progressive knowledge-led curriculum. *Asia-Pacific Journal of Teacher Education*, 43 (3), pp.210–224.

Lambert, D. and Morgan, J., 2010. *Teaching Geography 11–18: A Conceptual Approach*. Maidenhead: Open University Press.

Lefebvre, H., 1991. *The Production of Space* (Trans. D. Nicholson-Smith). London: Blackwell Publishing.

Livingstone, S. and Bovill, M., 2002. *Young People and New Media: Childhood and the Changing Media Environment*. London: Sage.

Livingstone, S. and Smith, P.K., 2014. Annual research review: Harms experienced by child users of online and mobile technologies: The nature, prevalence and management of sexual and aggressive risks in the digital age. *Journal of Child Psychology and Psychiatry*, 55 (6), pp.635–654.

Matthews, H. and Limb, M., 1999. Defining an agenda for the geography of children: Review and prospect. *Progress in Human Geography*, 23 (1), pp.61–90.

McKendrick, J., 2009. Localities: A holistic frame of reference for appraising social justice in children's lives. In J. Qvortrup, W. Corsaro and M. Honig (eds), *The Palgrave Handbook of Childhood Studies*. Basingstoke: Palgrave Macmillan, pp.238–255.

Ofcom, 2017. *Children and Parents: Media Use and Attitudes Report*. London: Ofcom. Available at: www.ofcom.org.uk/research-and-data/media-literacy-research/childrens/children-parents-2017 [Accessed on 14 February 2019].

Peet, R., 2015. Marxist geography. In D.J. Wright (ed.), *International Encyclopedia of the Social and Behavioral Sciences* (2nd edn). London: Elsevier, pp.674–687.

Plowman, L., Stephen, C. and McPake, J., 2010. *Growing up with Technology: Young Children Learning in a Digital World*. London: Routledge.

Reiss, M. and White, J., 2013. *An Aims-Based Curriculum: The Significance of Human Flourishing for Schools*. London: Institute of Education Press.

Roberts, M., 2017. Geographical knowledge is powerful if.... *Teaching Geography*, 42 (1), pp.6–9.

Skelton, T., 2008. Research with children and young people: Exploring the tensions between ethics, competence and participation. *Children's Geographies*, 6 (1), pp.21–36.

Tani, S., 2011. Is there a place for young people in the geography curriculum? Analysis of the aims and contents of the Finnish comprehensive school curriculum. *Nordidactica – Journal of Humanities and Social Science Education*, 1 (1), pp.26–39.

Valentine, G., 1996. Angels and devils: Moral landscapes of childhood. *Environment and Planning D: Society and Space*, 14 (5), pp.581–599.

Valentine, G., Skelton, T. and Chambers, D., 1998. Cool places: An introduction to youth and youth cultures. In T. Skelton and G. Valentine (eds), *Cool Places: Geographies of Youth Cultures*. London: Routledge, pp.1–34.

van Blerk, L., Barker, J., Ansell, N., Smith, F. and Kesby, M., 2009. Researching children's geographies. In L. van Blerk and M. Kesby (eds), *Doing Children's Geographies: Methodological Issues in Research with Young People*. Abingdon: Routledge, pp.1–8.

Van Dijck, J., 2013. *The Culture of Connectivity: A Critical History of Social Media*. Oxford: Oxford University Press.

Walshe, N. and Healy, G., 2020. *Geography Education in the Digital World*. London: Routledge.

Yarwood, R. and Tyrell, N., 2012. Why children's geographies? *Geography*, 97 (3), pp.123–128.
Young, M., Lambert, D., Roberts, C. and Roberts, M., 2014. *Knowledge and the Future School*. London: Bloomsbury Academic.
Young People's Geographies Project, 2011. *Young People's Geographies*. Available at: www.young-peoples-geographies.co.uk/ [Accessed 2 September 2019].

Part II

Geographical sources and connections in the digital world

Chapter 6

Geographical sources in the digital world
Disinformation, representation and reliability

Margaret Roberts

Introduction

In 2001, towards the end of the University of Sheffield Secondary Geography PGCE course, I asked students to share their experience of using computers in schools that year. They then voted on what they considered to be the most important developments. They had used PowerPoint (available only since 1998), data processing packages and computer disks holding massive amounts of information. Some students, including Richard Allaway, opted for the internet, which surprised many of us as this was not a discrete software package. His choice turned out to be prescient. Writing about the impact of information technology seventeen years later, Parkinson (2018) concurred, stating 'it was the arrival of the internet which had the most significant impact on schooling and geography classrooms' (p.185).

It is now 30 years since Tim Berners-Lee, a British engineer working for CERN (Conseil Europeen pour la Recherche Nucleaire), developed the system that was later to become the World Wide Web (WWW). His aim was to produce a powerful, global collection of documents and resources that could be linked to each other. In 1993 CERN allowed the WWW to be used on a royalty-free basis. The expansion of the web since then has revolutionised the availability of information worldwide and enabled those involved in geographical education to have access to vast amounts of information in various forms: text, statistics, maps, graphs, images and film. Bartlett and Miller (2011) proposed that:

> The internet is now almost certainly the greatest source of information for people living in the UK today ... The information we access and consume on the internet is central to forming our attitudes, our beliefs, our views about the world around us and our sense of who we are within it. (p.3)

This chapter identifies ways in which resources available now on the internet for the teaching and learning of geography are different from what was available previously. It explores opportunities and challenges the internet provides for both teachers and learners of geography. It refers to the notion of 'digital competence' needed to use the internet effectively.

Resources available before the use of computers in schools

Before there was access to computers, the main sources of information used in geography classrooms were sets of textbooks and atlases. These were supplemented by Ordnance Survey maps, wall maps, slides, film-strips, overhead transparencies, photo-packs, video-cassettes and sometimes by artefacts such as rock specimens. Local Authority advisors, the

Geographical Association (GA) and the Royal Geographical Society with Institute of British Geographers (RGS-IBG) contributed additional resources through lectures, conferences, local meetings and publications. They shared some common characteristics. First, although there was a choice of textbooks, atlases and visual materials, the amount was finite. Second, they were produced by people with expertise in geographical education. Third, information in them became out-of-date. Fourth, they had to be purchased out of departmental budgets.

The advent of computers in schools

As computers were gradually introduced into schools in the 1980s and 1990s, the range of resources produced for school geography increased. These included data handling packages and disks containing images, statistics, maps and text, some of which included software to enable users to analyse and present the information in other formats such as graphs and maps. These resources shared the characteristics of what was available earlier: there was a finite number of software packages; people with expertise in geographical education provided advice on how to use them; they became out-of-date; and some were very costly. The ability of school geography departments to access these computer-based resources, however, was limited by the small number of computers in schools and by varied access to the internet. The number of computers in schools increased after the Labour Government (1997–2010) made the use of ICT in schools a national priority and provided funding for computers, interactive whiteboards and training.

The arrival of the World Wide Web

Although the WWW was launched in 1993, its significance for geographical education was not immediately evident. Throughout the 1990s articles on information technology in the Geographical Association's journal for geography teachers, *Teaching Geography*, focused on hardware, software, GIS, getting connected and training. The first articles to focus on the web appeared in 1996 (Durbin and Sanders, 1996) and 1997 (Taylor, 1997). After 2000, use of the web in geographical education increased rapidly. The characteristics of online resources are significantly different from what was available before: the amount of information is seemingly infinite; most is not mediated for educational use; it can be up-to-date; and much is freely available. There are advantages and disadvantages of each of these characteristics.

The amount of information available can be illustrated by a Google search on a commonly studied topic, 'coastal erosion in Norfolk'. My laptop yielded an amazing 648,000 hits in 0.31 seconds. These included reports from the Geological Society, the North Norfolk District Council, newspaper articles, images and maps. Narrowing the search to 'coastal erosion at Happisburgh' and selecting 'videos' yielded 678 hits in 0.22 seconds, which included films on YouTube (a site available since 2005). Having quick access to relevant information when needed is clearly an advantage. Yet the sheer volume can be daunting.

It is possible to find, online, an impressive amount of resources produced by those with expertise in geographical education. The GA, the RGS-IBG, the BBC and examination awarding bodies all have their own websites. In addition, resources can be found on websites produced by geography teachers. For example, Matt Podbury, who teaches at the International School of Toulouse, is author of the website 'geographypods'.[1] Richard Allaway has made available, through subscription, all the resources used for geography courses at the

International School of Geneva ('geographyalltheway'[2]). Mitchell (2020) suggests that whilst such websites sit beyond large organisations 'teachers portray that these "unofficial sources" are trusted, used regularly and speaking to the geography teacher community in practical ways' (p.167). Nevertheless, only a tiny fraction of what is available online has been produced by people involved in education. It is possible for any organisation and anyone with an idea to set up a website and post information on it. Although some of this information can be useful for geography, it can also be inaccurate, biased, or misleading. This will be discussed later in the chapter.

Information on the web can be astonishingly up-to-date. We can see satellite images and weather maps representing the current situation. It is possible to find information about earthquakes that have occurred in the last hour from the US Geological Survey (USGS) Earthquakes Hazard programme. Newspapers and television news programmes report occurrences such as earthquakes, volcanic eruptions, floods and cyclones as they are happening. Online maps can be more up-to-date than offline maps, showing changes to place names, boundaries, routes, settlements and other features. Google Earth is updated about once a month, although not every image is updated so it is never completely accurate. On average the images it uses are between one and three years old. There are inevitably some time-lags related to statistics, depending on when they were collected. For instance, the most up-to-date UK census data is from 2011. The United Nations, the World Health Organisation, the Refugee Council, the World Bank and GapMinder are among organisations that regularly update their online statistics. Information from official reports is usually available online immediately on publication. For instance, on the day UNESCO's report on biodiversity was published, it was possible to access a press release (IPBES, 2019) with a link to a summary of the report and other resources including photographs, videos, raw video clips, and animations. There are clear advantages of having up-to-date information. It can be motivating for students to study events as they are actually happening. Also, out-of-date statistics give misleading views of other places, particularly in areas where there have been rapid changes, e.g. in life expectancy. The world and what we know about it is constantly changing. As geography is a dynamic subject, keeping up-to-date with new developments in the subject creates more demands on geography teachers. Mitchell (2020) and Puttick (2021) add that because this flow of information comes from beyond the discipline of geography, a further burden is placed on geography teachers to recontextualise new ideas from material available to them (e.g. news items).

Now, thanks to Bernars-Lee's vision, much of what is on the web can be accessed free of charge providing geography education with a bonanza of invaluable resources that previously had to be bought. Full access to some online resources such as Geographical Association journals, academic journals, some teacher produced websites and some newspapers (e.g. *Daily Telegraph, Wall Street Journal*) is available only through membership of organisations or subscription. Although many online resources are free, their use depends on hardware and connectivity, both of which involve costs in outlay and in maintenance. A BECTA report in 2006 suggested 'considerable variation within and across schools with regard to regular access to reliable technologies' (BECTA, 2006, p.3). In spite of an increase in the number of computers in schools, this statement is, from informal investigations I have carried out, still true today. I found that although all geography classrooms had teacher access to the internet, there were inequalities in access to additional hardware. For example, many schools had laptops or computer suites, but these were for shared used and therefore often not available for use in geography lessons. Many geography departments gave homework which required internet access. Not all schools assumed that all students had such access, so some made provision for them to use school equipment before and after school hours. Policies on use of smartphones in the classroom varied. Some schools had a total ban on smartphones. Some geography departments allowed students to use smartphones to access the internet on rare occasions under the instruction of the teacher. Some departments encouraged their use. The international and independent schools I contacted had, in contrast, one-to-one provision of internet access for students in all lessons, achieved through a policy of BYOD (bring your own device). This meant that the web could be accessed by students regularly and flexibly whenever it was appropriate in lessons.

Geographical sources in the digital world

Resources on the WWW are different from those that were available previously, not only because of the vast amount of up-to-date information freely available, but also because many online resources are fundamentally different. The web gives access to the kinds of data used by geographers in constructing geographical knowledge. For example, the Greater London Authority has made it possible, on its London DATASTORE website, to access and analyse primary data from the 2011 UK census (GLA, 2019). Although the Intergovernmental Panel on Climate Change (IPCC) does not carry out its own research, the secondary data presented in its reports are based on primary quantitative data collected by a range of scientists. The conclusions reached by the IPCC on climate change are based on reliable and consistent evidence. Similarly, secondary data published by the European Environment Agency on glacial retreat in Europe (EEA, 2016) is based on primary data collected by meteorologists and glaciologists. This data is used to produce well-evidenced conclusions.

The web gives access to more varied perspectives than were available previously. YouTube films, online photographs, articles and newspapers, produced by people from countries all over the world, can transform and enrich the way we see and understand places and help us guard against stereotyping and othering. The web appears to enable previously neglected groups to have a voice (Livingstone and Bovill, 2002). The web also gives access to 'cultural and artistic approaches to representing place', including art, poetry and music, now studied in A-Level geography (DfE, 2014, p.10). All these online representations, however, must be open to analysis, evaluation and critical scrutiny, as they could be partial or misleading. They might be partial because they have been produced for specific purposes (e.g. to promote tourism, to attract business or to argue for funds for regeneration or charitable purposes). It is worth probing what has been included and excluded and why. Individual voices, accounts and anecdotes can provide valuable perspectives, but they only ever give a view from a certain standpoint in time and space; therefore, it is important to consider their purpose and limitations for and within the geography classroom. Representations can be inadvertently misleading. For example, online case studies using Bangladesh as an example are mainly focused on flooding and overpopulation. In a professional development activity I frequently use, geography teachers always significantly underestimate life expectancy figures for Bangladesh. Their explanations for their out-of-date 'guesses' reveal negative perceptions of the country, which could be influenced by the legacy of these case studies. There is a risk of a case study promoting a 'single story' of a place (Adichie, 2009).

Rapid advances in technology since 2000 have enabled the production of applications that can be used interactively. Maps and satellite images on Google Maps and Google Earth, available since 2005, can be manipulated; they can be seen at different scales and particular features highlighted. The advancement of communications means that geography teachers can virtually communicate (through Skype) with a research scientist working in the Arctic (British Antarctic Survey, 2015). It is possible, using GapMinder (launched in 2005), to investigate selected indices of development for most countries in the world and to use the data to produce graphs for analysis. Developments in digital technology and mapping techniques have enabled the construction of the cartograms presented in WorldMapper, and in turn have 'changed the way we can understand the complexity of the world with its diverse social and physical dimensions' (Hennig, 2019, p.71). It is now possible, with these and other applications, to manipulate maps and graphs quickly, which leaves more time for making sense of them through classroom discussion.

Opportunities for teachers and learners

The web provides previously unimaginable opportunities for teachers. It is clearly an advantage to be able to access up-to-date resources and case studies relevant to the local area and to what is being studied. It makes sense for teachers to work collaboratively in searching for and selecting online resources, in exploring the potential of websites suggested in professional publications and in adapting online examples of schemes of work and lesson plans. This could be achieved through collaboration within geography departments, across groups of schools, e.g. in multi-academy trusts (MATs), or through online professional communities through social media platforms (e.g. Facebook and Twitter). Initial teacher education courses could promote collaborative investigations amongst their students and partner schools. This kind of collaboration can also involve stakeholders from beyond the school environment, and is not limited to just using online resources already available. For example, Rackley (2019) has highlighted how climate services, defined as the provision of 'climate information to help individuals and organizations make climate-smart decisions' (WMO, n.d.), can be used by geography teachers and students. More recently, Rackley has led the World Energy and Meteorology Council's (WEMC) Education Special Interest Group, involving geography teachers, climate/energy experts and policy makers, as they collaboratively design a new demonstrator for climate and energy data for school use. The Geo-Mentors network, which was jointly set up by Esri UK and the RGS-IBG has also allowed geography teachers to develop connections with industry experts, which has in turn shaped the online resources, including GIS data that they can go on to access with their students (Healy and Walshe, 2020).

The literature on the use of ICT in schools has identified various opportunities it offers for learners. BECTA referred to pupils having 'greater autonomy in their investigations' (BECTA, 2004) and identified the key advantage of ICT being the 'opportunity it offers for developing a more personalised learning environment' (BECTA, 2006, p.57) targeted at the needs of individuals. Hague and Payton (2010) envisaged students becoming increasingly independent in their use of the internet, 'as they are supported to find and select information for themselves' (p.54). Cambridge Assessment (2017) stated that digital technologies encouraged 'active learning, knowledge construction, inquiry and exploration on the part of the learners' (p.1). Redecker (2017) envisaged digital technologies being used 'to address learners' diverse learning needs, by allowing learners to advance at different levels ... and to follow individual learning pathways' (p.22).

Are these opportunities being taken? Has there been a shift towards more learner-centred pedagogic approaches? A NESTA (2012) report, which refers to all the opportunities mentioned above, stated: 'no technology has an impact on learning in its own right; rather its impact depends on the way in which it is used' (p.9). The way the web is currently used in geography varies considerably and has been influenced by access to hardware in school and by approaches to pedagogy usually adopted. The range of approaches can be considered along a 'participation dimension' (Table 6.1), a framework I have used in relation to geographical enquiry (Roberts, 2013). In the context of this chapter, the categories 'controlled', 'framed' and 'negotiated' indicate the extent to which students are actively involved in the use of the web. In geography, progression in the use of the web can be considered in terms of an increase in the conceptual demands of what is being studied, the complexity of the online resources needed to investigate it and the extent to which students are dependent on teachers' support in using them.

In practice, a scheme of work, lesson or an activity based on the use of the web might include all parts of the participation dimension, as illustrated in the example (Figure 6.1). Students carried out this activity after studying Hurricane Harvey and being introduced to key concepts of causes and consequences, and social, environmental, economic and political

Table 6.1 The participation dimension.

Stage of investigation	Controlled	Framed	Negotiated
The need to know: focus of study	Teacher explains focus of study 　　Teacher chooses website(s)	Students use list of websites provided and search for data to investigate a geographical question or issue	Students search the internet to find websites and resources to investigate a geographical question or issue
Using data	Teacher selects and presents data on interactive whiteboard	Students are given frameworks to guide selection of data	Students analyse and interpret data
Making sense	Students listen, watch, answer questions and carry out activities	Students carry out activities requiring engagement with the data, e.g. analysis, interpretation, comparisons or reconstruction to present in a different format, e.g. PowerPoint	Students use data to develop arguments and reach their own conclusions. They decide how to present their findings
Reaching conclusions	Students expected to reach anticipated conclusions	Students discuss conclusions	Students reach their own conclusions
Comment	Teacher makes all decisions, keeping control of content and activities. Students do not access web during lessons	Enquiry-based 　　Teacher provides supportive scaffolding in instructions and through discussion to enable students to use web themselves	Students make their own decisions. Teachers monitor work, provide guidance and encourage critical evaluation of work
Teaching style	Transmission, used to 'cover' content	Enquiry-based, nominally encouraged by some GCSE and A-Level specifications.	Project-based coursework, required for Non-Exam Assessment at A-Level

> **Instructions**
>
> How was a specific hurricane caused? What damage was caused by a specific hurricane? Did the area ever recover from the specific hurricane?
>
> You are a reporter for National Geographic and you have been given the assignment to develop a newspaper report (750 word limit) that will describe and explain the impact a hurricane has had on an area. You can choose any hurricane that has occurred throughout the world; there is no constraint on location or year that it swirled.
>
> **Content**
> Headline
>
> - Location – Where did the devastation take place? (remember to include a map)
> - Description of what happened – Why did the hurricane take place? Time line of events. Include diagrams.
> - The effect of the damage – economic (buildings, roads, infrastructure), social (people forced to move), and environmental (landscape). Try and include eye witness accounts.
>
> What is being done to rebuild the damage caused? How can we inform people about the danger of hurricanes? E.g. evacuation/emergency plans

Figure 6.1 Illustration of how a lesson activity can include all parts of the participant dimension from the 'Storm Chasers' lesson.
Source: Extracts from website Geogalot,[3] set up by Ellena Mart, International School of Geneva.

factors, which modelled a geographical way of studying hurricanes. In this web-based investigation, key questions were controlled by the teacher. The teacher provided a supportive framework via the enquiry questions, a list of 16 different websites relate to five different hurricanes, advice on what might be included in the newspaper report and by making the assessment criteria clear. There were opportunities for students to negotiate their own choice of hurricane.

Challenges for teachers and learners

All geographical knowledge is provisional, contestable and open to critical scrutiny regardless of whether it is accessed offline or online. The discipline of geography has been shaped by the kinds of questions asked by geographers and these are influenced by the concerns and cultural contexts of the time. In England, what has been included in the geography national curriculum and content requirements for public examinations has been influenced by the views of different governments (Rawling, 2001; 2015). Working within government guidelines, geography textbook authors and those who devise examination specifications make further decisions about content. This constructivist view of the curriculum does not mean that anything goes. Geography educators who produce resources set out to provide information and explanations that are accurate and reliable and present authentic representations of the world. But representations of the world are not neutral: 'Our knowledge of the world is always from a certain standpoint, a certain location. We see it from here rather than from there' (Allen and Massey, 1995, p.2). Even before the arrival of the web, there has been a need to develop critical awareness about how knowledge about the world has been constructed and filtered for classroom use and ways in which it might be biased or misleading.

Online resources present additional challenges. In what has been termed 'post-truth' times, how can we know whether what is found online is true or false? How can we know whether it is based on reliable evidence? How can we know whether online images have been manipulated? Is online information 'fake'? In recent years there has been increasing use of the term 'fake news', but definitions and usage vary. The term is defined by Collins English Dictionary (2019) as 'false, often sensational, information disseminated under the guise of news reporting'. A Reuters Institute (2017) report identified three categories of fake news: '1) news that is invented to make money or discredit others; 2) news that has a basis in fact, but is 'spun' to suit a particular agenda, and 3) news that people don't feel comfortable about or don't agree with' (p.20). A European Commission (2018) report favoured the term 'disinformation' over 'fake news' and defined it as 'all forms of false, inaccurate or misleading information designed, presented and promoted to intentionally cause public harm or for profit' (p.10). It needs to be recognised that there is no gatekeeper to check the accuracy and authenticity of what is included on websites. It is likely that teachers and students of geography will encounter two types of disinformation in their online searches: firstly, information that is inaccurate and verifiably false, and secondly, information that has a basis in fact but is slanted to promote or support a particular agenda.

We need to be able to trust that online resources are reliable; accurate data provide the necessary backing for well-developed arguments and the knowledge claims made in geography. Students should develop knowledge of websites that provide reliable information and become aware of why this is so. The websites of many organisations including the Meteorological Office, the United Nations, the European Environment Agency, the World Health Organisation, USGS, NASA and the Intergovernmental Panel on Climate Change are reliable. They identify the sources of the data used and draw on the work of academic researchers which has been scrutinised by peer review for accuracy and reliability before publication.

Much information that teachers and students are likely to encounter from online searches is on partisan websites that promote particular agendas. Partisan websites usually include accurate information, but it is selected and presented to support specific interests. Websites of many Think Tanks, with seemingly neutral and authoritative names, promote political agendas. For example, the 'Institute for Economic Affairs' (IEA) is a high-profile right-wing UK think tank, founded in 1955, promoting free market solutions to a wide range of social and economic issues. The 'Institute for Public Policy Research' (IPPR) was set up in the UK in 1988 to promote left-wing thinking. The USA's 'Heritage Foundation' was set up in 1973 to advance principles of a conservative agenda of free enterprise and limited government. The fact that what is included on Think Tank websites is biased does not mean that such websites should be avoided. On the contrary, it is important to develop an understanding of the ideologies which influence decisions that have geographical implications and to become aware of the assumptions underpinning what is included on their websites.

Partisan websites can be useful in geography for the investigation of controversial issues. For example, different information and arguments related to the proposed expansion of Heathrow Airport are presented on the websites of: Heathrow Airport Holdings with its commercial interests; the UK government, which gave approval for expansion; Greenpeace and Friends of the Earth, both of which oppose expansion; and local and national newspapers. There is no longer any justification for inventing 'fake' quotations about controversial issues when it is easy to find actual views online. Although journalists set out to provide accurate information, the ways in which newspapers present issues such as Heathrow expansion vary considerably. Influenced by their underpinning values, they select different factual pieces of information and give emphasis to different arguments. To use only one partisan website to investigate a topic is risky; it leads to a partial and biased understanding of a geographical issue. To use several partisan sites with different views can, however, if students learn to probe their underpinning values and assumptions, contribute to the development of a critical understanding of geographical issues.

Another way in which online searches can mislead is related to the algorithms which search engines use to list the results of a search. Organisations and individuals producing websites can use various strategies to achieve a higher ranking on search engine pages, including search engine optimisation (SEO) – which adjusts website content and design to achieve higher ranking – and search engine marketing (SEM) – which increases visibility mainly through paid advertising. So the rank order of listings is not neutral. Furthermore, the order of listings is influenced by interests indicated by the previous searches of the user. Ultimately, search engine algorithms play a role in shaping what information is likely to be accessed and used by geography teachers in their curriculum making (Mitchell, 2020).

If teachers and students are to make the best use of the web in geography, they need to be aware of the opportunities and challenges discussed above and to develop a range of competences. When computers were first introduced into schools, there was emphasis on developing the technical skills required to use them (BECTA, 2006). It has since been recognised that a broader range of skills is needed. Bartlett and Miller (2011) used the term 'digital fluency' to describe 'the ability to find and critically evaluate online information' (p. 4). The terms 'digital literacy' and 'digital competence' are commonly applied to the use of all aspects of ICT. The European Commission has produced a framework, 'DigComp 2.0', (Vuorikari et al., 2016) in which five key components of digital competence are identified, one of which, 'Information and Data Literacy' (Table 6.2), is highly relevant to this chapter. It demands more than technical skills, analysis and interpretation. The DigComp 2.0 framework emphasises the need for critical evaluation, not only of the data and information themselves, but also of the 'credibility and reliability' of their sources. In this digital age,

Table 6.2 Extract from the EU's DigComp 2.0: The Digital Competency framework for Citizens. Component 1: Information and Data Literacy.

Competence areas	*Competences*
1.1 Browsing, searching and filtering data, information and digital content	To articulate information needs, to search for data, information and content in digital environments, to access them and to navigate between them. To create and update personal search strategies
1.2 Evaluating data, information and digital content	To analyse, compare and critically evaluate the credibility and reliability of sources of data, information and digital content. To analyse, interpret and critically evaluate the data, information and digital content
1.3 Managing data, information and digital content	To organise, store and retrieve data, information and content in digital environments. To organise and process them in a structured environment

Source: Vuorikari et al. (2016, p.8).

when we encounter vast amounts of information and various kinds of disinformation on the internet, these critical competences are crucially important for geographical education. Geography teachers can contribute to the development of digital competence from Key Stage 1 (children of age 5) onwards, but only if students use the web themselves in their lessons. If students are to develop these competences they need increasing opportunities to make their own decisions about websites, selection of data, analysis and presentation and, most importantly, opportunities to discuss and reflect critically on these decisions.

Final thoughts

The internet, with its wealth of easily accessible free resources, has been significant for geographical education. Much of the information used directly by teachers and students, and indirectly by authors of textbooks and other offline resources, comes from the internet. In school, students develop their knowledge and understanding of the world mainly through how geography is represented to them in these resources. As information from the web is so pervasive, we have a responsibility to use it critically to develop students' competences in using it effectively too. Yet, partly because of inequalities in access to hardware, financial constraints and pressure to 'cover' the curriculum, the opportunities in some schools for students to become effective users of the web in geography are restricted. We need to continue to question the use of the internet within geographical education: What is used? When is it used? Who uses it and in which schools? How is it used? Why is it used in this way? How does it contribute to students' geographical education? We need to develop an understanding of the ways in which use of the internet in geography can empower students, so that they can use this incredibly powerful source of information confidently and critically to make sense of the world not only in the classroom, but also in their future lives.

Notes

1 www.geographypods.com
2 www.geographyalltheway.com
3 www.geogalot.com

References

Adichie, C.N., 2009. *The Danger of a Single Story*. Available at: www.ted.com/talks/chimamanda_ngozi_adichie_the_danger_of_a_single_story [Accessed on 26 December 2019].

Allen, J. and Massey, D. (eds), 1995. *The Shape of the World: Explorations in Human Geography*. Oxford: Oxford University Press.

Bartlett, J. and Miller, C., 2011. *Truth, Lies and the Internet: A Report into Young People's Digital Fluency*. London: Demos. Available at: www.demos.co.uk/files/Truth_-_web.pdf [Accessed 31 May 2019].

BECTA, 2004. *What the Research Says about Using ICT in Geography*. Coventry: BECTA. Available at: https://dera.ioe.ac.uk/1653/66/wtrs_geography_Redacted.pdf [Accessed 31 May 2019].

BECTA, 2006. *The Impact of ICT in Schools: A Landscape Review*. Coventry: BECTA. Available at: https://webarchive.nationalarchives.gov.uk/20101017100205/http:/research.becta.org.uk/upload-dir/downloads/page_documents/research/impact_ict_schools.pdf [Accessed 31 May 2019].

British Antarctic Survey, 2015. *News Story: Live link to the Arctic*. Available at: www.bas.ac.uk/media-post/digital-explorer-launches-virtual-adventure-on-the-ice-at-the-uk-arctic-research-station/ [Accessed 19 May 2020].

Cambridge Assessment, 2017. *Digital Technologies in the Classroom*. Cambridge: UCLES. Available at: www.cambridgeinternational.org/Images/271191-digital-technologies-in-the-classroom.pdf [Accessed 31 May 2019].

Collins English Dictionary, 2019. Fake news. In *Collins English Dictionary*. London: Harper Collins. Available at: www.collinsdictionary.com/dictionary/english/fake-news [Accessed 31 May 2019].

Department for Education (DfE), 2014. *Geography GCE AS and A Level Subject Content*. London: DfE.

Durbin, C. and Sanders, R., 1996. Geographers on the internet. *Teaching Geography*, 21 (1), pp.15–19.

European Commission, 2018. *A Multi-Dimensional Approach to Disinformation: Report of the Independent High Level Group on Fake News and Online Disinformation*. Available at: https://ec.europa.eu/digital-single-market/en/news/final-report-high-level-expert-group-fake-news-and-online-disinformation [Accessed 31 May 2019].

European Environment Agency (EEA), 2016. *Glaciers*. Available at: www.eea.europa.eu/data-and-maps/indicators/glaciers-2/assessment [Accessed 31 May 2019].

Greater London Authority (GLA), 2019. *LondonDATASTORE London*. Available at: https://data.london.gov.uk/census/ [Accessed 31 May 2019].

Hague, C. and Payton, S., 2010. *Digital Literacy across the Curriculum*. Futurelab. Available at: www.nfer.ac.uk/publications/FUTL06/FUTL06.pdf [Accessed 31 May 2019].

Healy, G. and Walshe, N., 2020. Real-world geographers and geography students using GIS: Relevance, everyday applications and the development of geographical knowledge. *International Research in Geographical and Environmental Education*, 29 (2), pp.178–196. doi: doi:10.1080/10382046.2019.1661125.

Hennig, B., 2019. Remapping geography: Using cartograms to change our view of the world. *Geography*, 104 (2), pp.71–79.

Intergovernmental Science-Policy Platform on Biodiversity and Ecosystem Services (IPBES), 2019. *IPBES Global Assessment Underscores Need for Transformational Change to Safeguard Life on Earth*. Available at: www.cbd.int/doc/press/2019/pr-2019-05-06-IPBES-en.pdf [Accessed 31 May 2019].

Livingstone, S. and Bovill, M., 2002. *Young People and New Media: Childhood and the Changing Media Environment*. London: Sage.

Mitchell, D., 2020. *Hyper-Socialised: How Teachers Enact the Geography Curriculum in Late Capitalism*. London: Routledge.

NESTA, 2012. *Decoding Learning: The Proof, Promise and Potential of Digital Education*. London: NESTA. Available at: https://media.nesta.org.uk/documents/decoding_learning_report.pdf [Accessed 31 May 2019].

Parkinson, A., 2018. The impact of technology on geography and geography teachers. In M. Jones and D. Lambert (eds), *Debates in Geography Education* (2nd edn). London: Routledge, pp.184–196.

Puttick, S., 2021. Digital technologies and their roles in knowledge recontextualisation and curriculum making. In N. Walshe and G. Healy (eds), *Geography Education in the Digital World*. London: Routledge, pp.17–28.

Rackley, K., 2019. Resources to teach the changing nature of climate and energy. *Teaching Geography*, 44 (2), pp.62–65.

Rawling, E., 2001. *Changing the Subject: The Impact of National Policy on School Geography 1980–2000*. Sheffield: Geographical Association.

Rawling, E., 2015. Curriculum change and examination reform 14–18. *Geography*, 100 (3), pp.164–168.

Redecker, C., 2017. *European Framework for the Digital Competence of Educators*. Luxembourg: European Union. Available at: https://ec.europa.eu/jrc/en/publication/eur-scientific-and-technical-research-reports/european-framework-digital-competence-educators-digcompedu [Accessed 31 May 2019].

Reuters Institute, 2017. *Digital News Report 2017*. Oxford: Reuters Institute and Oxford University Press. Available at: https://reutersinstitute.politics.ox.ac.uk/sites/default/files/Digital%20News%20Report%202017%20web_0.pdf [Accessed 31 May 2019].

Roberts, M., 2013. *Geography through Enquiry: Approaches to Teaching and Learning in the Secondary School*. Sheffield: Geographical Association.

Taylor, L., 1997. Using the World Wide Web as a geography resource. *Teaching Geography*, 22 (1), pp.11–15.

Vuorikari, R., Punie, Y., Carretero Gomez, S. and Van den Brande, G., 2016. *DigComp 2.0: The Digital Competence Framework for Citizens. Update Phase 1: The Conceptual Reference Model*. Luxembourg: European Union. Available at: https://publications.jrc.ec.europa.eu/repository/bitstream/JRC101254/jrc101254_digcomp%202.0%20the%20digital%20competence%20framework%20for%20citizens.%20update%20phase%201.pdf [Accessed 7 January 2020].

World Meteorological Organization (WMO), n.d. *What Are Climate Services?* Available at: www.wmo.int/gfcs/what-are-climate-services [Accessed on 26 December 2019].

Chapter 7

'Connecting the Classroom'
Teaching geographies of development via digital interactive spaces

Rory Padfield

Introduction

The two quotes below neatly summarise some of the key challenges that face the teaching of geography in higher education (HE):

> Today's teachers have to learn to communicate in the language and style of their students, this doesn't mean changing the meaning of what is important or of good thinking skills, instead it means to reconsider the methodology and content of education. (Clarke and Clarke, 2009, p.397)

> Geographers learn through the soles of their feet. (McEwan, 1996, p.379)

As Clarke and Clarke (2009) observe, educators must ensure that teaching methods and curriculum content are relevant to the latest generations entering universities and colleges. In view of the influence of the digital revolution on HE (Ferguson et al., 2017; Ferguson et al., 2019), there is clearly significant scope to incorporate innovations that respond in the most effective way. This includes techniques and approaches such as 'just-in-time' teaching, 'flipped classrooms', access to hyperlinked multimedia information, parallel processing and multi-tasking, and so forth (O'Hara, 2007; Pike, 2015). Geography is well-positioned to respond to the 'digital pedagogic challenge' and, to-date, has a commendable track record in incorporating innovative approaches in teaching and learning in both school and HE (De Miguel González and Donert, 2013).

The second quote by McEwan (1996) emphasises the importance of fieldwork in the study of geography. Typically, fieldwork involves students travelling off-campus to experience various human or physical geography related study techniques. In addition to gaining experience of the physical environment, fieldwork usually has the aim of bringing geographers into contact with the people and communities who live, work or traverse in that place. Yet not all subjects are in a position to take advantage of fieldwork in the same way. Geographies of development – broadly understood as the study of the interactions of society and environment, on uneven and unequal development, and the nature of local environments in the Global South (Potter et al., 2018) – is growing in popularity within geography HE programmes, yet without resources and time for overseas travel is largely taught as a conventional lecture-based module.

Thus, acknowledging the growing influence of digital technologies in HE pedagogy and the importance of fieldwork in a geography degree, are there ways to teach geographies of development that bring students into 'virtual' contact with people, communities and environments of the Global South? Can these interactions be beneficial for all parties involved in both the Global North and Global South? In view of these questions, this chapter reflects on a recently trialled learning experiment at Oxford Brookes University called 'Connecting the Classrooms' (CTC). Its aim was to facilitate a two-way, cross-cultural appreciation of a development geography topic – specifically, development geographies of Indonesia – between a group of UK and Indonesian students via an interactive digital platform. The centrepiece of the CTC was an online platform that facilitated 'virtual' interaction between both groups of students around a series of structured learning tasks. The chapter reviews the strengths, weaknesses and practicality of the experiment and offers reflections of the saliency of digital technology for the teaching of development geography.

Harnessing 'digital pedagogies' in higher education

In recent years, there is growing awareness of the vast potential for digital tools and techniques to support innovative teaching and learning methods in HE. A barometer of the growing role played by digital technologies is illustrated in the 2017 and 2019 'Innovative Pedagogy' reports led by Dr Rebecca Ferguson from the Open University. In the 2017 report there is a relatively limited discussion of the role played by digital technologies in the emerging trends in HE pedagogy (Ferguson et al., 2017). Yet in the 2019 report digital technologies take centre stage as a means to enhance teaching and learning approaches in HE. Examples include digital games, virtual studios, drone-based learning, and technology-enhanced assessment (Ferguson et al., 2019).

The application of digital tools and technologies has significant potential to contribute to a wide range of learning outcomes in geography, including innovative ways to engage learners from different backgrounds and places on topics of mutual interest. Two examples are drawn upon below with reference made to related academic theories and approaches. The first example is a case study elaborated in the 2017 'Innovative Pedagogy' report called the Humans of New York (HONY) project (Ferguson et al., 2019). HONY's centrepiece is a blog that introduces street portraits and interviews by the site's author, which has gathered more than 20 million followers on Facebook and Instagram. Followers can offer their views and responses to the images and stories and are encouraged to share similar or contrasting experiences. In reflecting on the responses to the HONY project, Ferguson et al. (2019, p.23) argue that:

> … this approach can promote constructive contact between people with various cultural backgrounds. A study of the project found many expressions of empathic resonance, reasoning, and response in the comments and discussion.

While not an example drawn specifically from HE practice, the HONY project demonstrates the potential to harness digital technologies as means to 'connect' different groups of learners on a topic of mutual interest. This is particularly the case where the different groups of users – in this case Facebook and Instagram users – demonstrate expressions of empathy and reasoning towards the experience and perspective of others. This particular outcome resonates strongly with ideals of the internationalisation of curriculum (IoC) concept. IoC refers to an equitable learning curriculum and experience that values social inclusion, cultural pluralism and global citizenship ahead of partisan links with any smaller geographical, cultural or social unit. Adapting earlier work by Leask (2001), geographer Martin Haigh (2002, p.53) identifies six indicators of IoC:

- Open to 'Otherness';
- International perspective;
- Self-aware;
- Aware of international professional contexts;
- Respect 'Otherness'; and
- Understands 'Otherness'.

Kitano (1997) takes an instructive approach to IoC by describing three levels in which a curriculum can incorporate international education. The first level is referred to as 'exclusive' and involves teaching in a traditional way (i.e. didactic lecture style); the lecturer behaves as the grand instructor or purveyor of knowledge which results in a predominantly conventional learning experience for the students. The second is called 'inclusive' and involves a traditional view but with alternative perspectives incorporated at various junctures. Such alternatives include a diversity of readings, guest lecturers and other resources (i.e. technology). The instructor employs a range of different teaching methods and various assessment types. The third is called 'transformative' and is based on a number of components, such as challenging traditional perspectives and assumptions, analysis of content through non-dominant perspectives, and instructional approaches focused on student knowledge and experience. Critical thinking is also a component within Kitano's approach. As opposed to exclusive and inclusive learning, students are equal participants in the learning process (Kitano, 1997). The HONY project has elements of 'transformative' learning, since it allows users to challenge traditional perspectives and builds in instructional learning through the user's own knowledge and experience.

The second example is the 'Global Classroom' run by Drexel University and University of Leeds, referred to as DLGC (Ferris and Wilder, 2017). The aim of the DLGC was to provide students with opportunities for international experiences while remaining on campus. It consisted of a four to six week series of activities, in which the students from the respective universities worked together remotely to create a new product and/or service, which they had to pitch (simultaneously) to a panel of experts and academics in both the United Kingdom and United States. Overall, the initiative is shown to have a number of positive outcomes, not only for the participating students but also for staff and the university as a whole. For example, student outcomes include working collaboratively, learning new techniques, and experiencing 'studying abroad' without leaving the classroom. As with the HONY example, the DLGC project engages with principles of IoC, offering students international perspectives, appreciation of working with learners from different backgrounds (i.e. 'the Other') and emphasising 'transformative' as opposed to 'exclusive' and 'inclusive' learning approaches. This chapter will now turn to the trialled learning experiment, CTC, at Oxford Brookes University.

'Connecting the Classroom': A case study in practice

This section describes the specific tasks and activities of the CTC exercise as run between January and May 2018. The aim of the exercise was to facilitate experiential learning and a cross-cultural appreciation of development geographies of Indonesia between a group of 82 undergraduate students in the Department of Social Sciences, Oxford Brookes University (OBU), and 35 undergraduate students in the Department of International Relations, Universitas Indonesia (UI).

Three factors required some thought before CTC could commence. First, the ethics of such an exercise needed careful consideration by the academic instructors before initiating any interaction with students. The main ethical issues identified in the context of CTC was the potential for unequal access to digital technology, including high-speed internet between the groups of students, and the need to set expectations for appropriate online behaviour. These were addressed by: i) understanding the respective digital accessibility challenges and making allowances for it in the planning of online-based activities; and ii) drafting a code of conduct for virtual engagement and circulating to students prior to any online activity (as discussed below). Second, it was important to identify a counterpart who could lead CTC in a university in the Global South. Ideally, the counterpart needed to be a module leader for a course covering a similar topic to the one taught in the UK, for example, geography, international relations, development studies or political science. Third, language can present a barrier to communication between the two groups of students. Ideally, both sets of students would have a good grasp of a common language to allow communication between the groups. In this case, the majority of the Indonesian students were proficient in English, although some students did feel anxious about communicating their views in written English. CTC was organised into the following five stages.

Stage 1: Pre-CTC set up

Early communication between the module leaders in both universities took place to ensure the overall aims of CTC, as well as the respective group projects (see Table 7.1 see p.69), aligned with the respective learning outcomes and assignments of modules in both universities. Module leaders discussed the aims and key issues of the project topics, as well as practical aspects related to the allocation of students into groups and timings. The specific project topics were selected in view of the key themes and topics covered in the respective modules. An appropriate online forum was identified for the purposes of online interaction between the two groups of students (Oxford Educational Cloud (OEC)).[1] A code of conduct for the online discussion was drafted in preparation for the first student meeting.

Stage 2: CTC introduction

Students in both universities were independently briefed by their respective module leaders of the overall project aims, objectives and how CTC feeds into the broader learning outcomes of the module. Students were allocated into groups of five or six and each group was assigned a research topic from Table 7.1. The code of conduct for online interaction was circulated to the students for their reference.

Stage 3: Research time and presentations by Oxford Brookes University students

In their allocated groups, students researched their topics with time set aside in lectures and tutorials for group working. Groups were expected to work between 1–2 hours per week on their projects outside of the lectures. OBU students prepared a ten minute oral presentation while the UI students documented their research findings in written form. After four weeks

Table 7.1 Project topic, background and key research questions for the 'Connecting the Classroom' exercise.

Project topic	Background and key questions for investigation
Tropical peatlands – a curse to development in the Global South?	The aim was to examine whether tropical peatlands – a soil type of considerable ecological value yet susceptible to fire and negative climate change impacts when developed in less sustainable ways – are a curse to development. Students addressed the following questions: i) Where are tropical peatlands found in the Global South and what are their different ecological, social and economic uses? ii) In what way do these uses conflict? iii) Ultimately, are tropical peatlands a curse to development or is there a more sustainable path to be found in the future?
Indonesia, deforestation and sustainable development	Indonesia has one of the largest areas of remaining tropical forest in Southeast Asia, yet has been subject to considerable deforestation in the past few years. Acknowledging the contrasting values that different groups of people place on the forests of Indonesia, students addressed the following questions: i) What are the causes and effects of deforestation in Indonesia? ii) Should forests be left alone on conservation grounds or developed for economic gain? iii) Is there a realistic 'middle ground' or alternative sustainable development model that can satisfy all people, including Indonesians and those living outside of the country?
Land grabbing: neo-colonialism or a genuine opportunity for development?	Large-scale land acquisition, often referred to as 'land grabbing', is a common characteristic of past and contemporary development in Indonesia. At the centre of this issue are questions related to the morality of land acquisition on this scale, especially in light of the growing food crisis in many parts of the Global South. Drawing on case studies from the Global South, students addressed the following questions: i) Where in the Global South has land grabbing taken place historically and more recently, i.e. in the last 10 years? ii) Which group of actors are typically in conflict over this issue? iii) On balance is land grabbing a form of neo-colonialism or a genuine opportunity for development?
Sustainable palm oil	Sustainable palm oil has come to represent a key debate in the contemporary Global South. We ask students to analyse the problem of expansion (land conversion causing deforestation) versus intensification (higher yields and extraction rates). We encourage students to debate the prospects for sustainability when economic growth, not biodiversity conservation, is the ultimate goal. Students are expected to address the following questions: i) How is sustainable palm oil broadly understood in current debates? ii) How should vulnerable smallholders respond to market pressures and conservationist concerns and what incentives are there for better land use management? iii) Ultimately, will environmental sustainability always be sacrificed at the expense of economic growth?

of group work, each of the OBU groups gave their presentations during one the lectures. Each presentation was filmed and feedback provided by the module leader. One presentation per topic was selected by the module leader and uploaded to the OEC.

Stage 4: Online interaction

The OEC website allowed students from both universities to view the videos and post written feedback on the website. The feedback function of the website operated in the same way as a typical social media platform, such as Facebook, whereby comments were posted sequentially underneath the videos. This allowed users to view the comments and any subsequent replies to comments made by the OBU students. This provided the opportunity for a form of online dialogue between the different groups of students. Each group from both universities was encouraged to post a minimum of one comment per video. Student comments were not moderated; instead, students were expected to comply with the code of conduct for online discussion, as circulated at the beginning of the exercise. Any breaches of the code of conduct would have resulted in expulsion from the exercise and the potential for escalation at a university level, if necessary. Overall, interactions between the two groups of students were generally constructive and supportive. On more than one occasion there were expressions of support for an opinion, as well as disagreement.

Stage 5: In-class reflection

Following the completion of online feedback, time was set aside in a lecture by both module leaders to reflect on the key learning outcomes from the CTC exercise. Each group was asked to report informally on their experience of the exercise, with emphasis placed on reflection on the way in which the dialogue with the other set of students had informed their understanding and perspective of the research topics.

Reflection

As illustrated in Figure 7.1 below, CTC has the potential to add significant value to the teaching of development geography. Drawing parallels with the Global Classroom project (Ferris and Wilder, 2017), CTC can be integrated into a programme in order to internationalise the curriculum. Conventional development geography courses have limited engagement or interaction with people from the Global South; perspectives of development

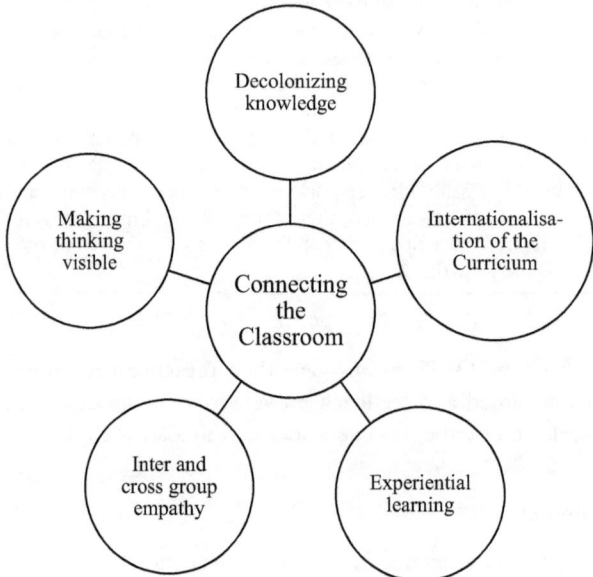

Figure 7.1 Learning outcomes of the 'Connecting the Classroom' exercise.

from people or communities from the Global South occur via documentaries or engagement with non-Western perspectives of development accompanied with reference to non-Western intellectuals or academics. On occasion, guest speakers from the Global South are invited to speak as part of a module. Acknowledging the difficulties of integrating international travel within a UK undergraduate degree programme, it is the expectation that development geography academic programmes are taught within the confines of lecture halls and classroom. Yet, in the case study described in this chapter, CTC gave the OBU students an opportunity to interact in a meaningful manner with those who have close or direct personal experiences of Indonesian development. These interactions were diverse, rich and not moderated. A testimony from one of the OBU students illustrates this point:

> The most beneficial aspect of the [CTC] process was reading the students' responses to our argument. This gave our group new perspectives on the issues surrounding tropical peatlands not readily available in journal articles.

Likewise, for the UI students there was an opportunity for insight into Western perceptions of development. In viewing the OBU filmed presentations and engaging in the online discussions UI students could reflect on their own assumptions and offer alternative ideas and perspectives. For instance, in a discussion about the relationship between neo-colonialism and land grabbing, a UI student suggested the OBU students should consider a different view:

> 'So, our point here is to take … a look at another face of neocolonialism that happened in the territory where land grabbing is now becoming [more of a] major trend in government policy'.

Aligning with notions of 'decolonizing knowledge' (Ferguson et al., 2017) and 'transformative learning' (Kitano, 1997), interactions between students helped both sets of students engage critically with their own positions and assumptions about development. Incorporating space and time for critical reflection supports ideas of experiential learning – 'learning through reflection and doing' (Felicia, 2011, p.1003) – whereby student-to-student interaction facilitates the development of their own knowledge and experience. Indeed, deeper engagement with the key concepts of development and reference to a more diverse literature within a number of the final assignments by OBU students – as compared with the previous year – suggests CTC had a positive outcome on the students' learning experience.

A further outcome of the exercise was the opportunity for the development of interpersonal social skills, specifically inter-group empathy. Ferguson et al. (2017) argue that 'virtual spaces' have a role in fostering empathy between different groups and communities. While the interactions between the two groups of students were never confrontational, there were occasionally differences of opinion. For the UI students, engaging with differing views via engagement with people's lived experience of a topic – in this case rural development in Indonesia – is likely to have contributed to an appreciation of and respect for the other group. Taking these points together, CTC thus addresses a number of Haigh's indicators of IoC, with explicit reference to openness and respect for 'Others', bringing an international perspective, and self-awareness (Haigh, 2002).

Relatedly, the online discussion and video download function facilitated 'making thinking visible'. Ferguson et al. (2019, p.5) argue that:

> ... learning becomes more effective when students can visualise their thinking ... as they do so, they leave traces of their thinking in the form of written marks and interactions with digital media, such as videos ... These visible records of students' personal and social learning can become resources for reflection. Teachers can see how each student is progressing towards mastery of a topic and can identify where students are blocked or have misunderstood a topic.

The CTC project facilitates the learning experience as students can track their own strengthening of knowledge and understanding: from the initial group presentation videos, to the webpage, through to the written dialogue between the groups of students. Moreover, since the content of the videos and dialogue are visible to those participating in CTC, learning can be achieved by all the students as they track and reflect on the progression of ideas and understanding communicated by individual students. In short, the visible nature of the learning outputs in CTC facilitates collective learning through the experience of individual students. The visible record of output is also useful from the perspective of the educator in that it allows targeted intervention and clarification where necessary.

To improve the students' learning experience of CTC, four recommendations are proposed. Firstly, opportunities to further develop the interaction between the students should be considered. In reflecting on the experience of CTC one of the OBU students observed:

> Overall, this exercise was insightful and beneficial to my learning. The only way it could be improved is by having more contact with the Indonesian students to learn more from their experiences to feed back into our own work.

One suggestion is to build in time for 'face-to-face' remote engagement between the students via video conferencing applications (e.g. Skype, WhatsApp, Zoom). Students are familiar with these digital applications and thus on a technological level would not present too much of a challenge. These interactions should be encouraged outside of scheduled lectures to make best use of the students' time whilst overcoming difficulties faced if there are significant time differences between the two countries. In keeping with the notion of 'making learning visible', video files should be recorded and uploaded to the CTC website as a means of tracking the progression and advancement of student understanding and knowledge.

Secondly, it is proposed that CTC ends with a video conference between the two groups of students. While such an activity may present challenges if there is a significant time difference between the two countries, a structured session with input from both groups would allow for reflective insights and reflections.

Thirdly, to help ensure both groups of students participate as equally as possible in the project, it is proposed that an assessed assignment might be incorporated into CTC. Such an assignment could include assessment of aspects of the students' digital outputs, i.e. video or dialogue material uploaded to the website, as well as a piece of summative reflective writing on how CTC has enhanced their knowledge.

Finally, to understand more explicitly the impact of CTC on the students' wider geographical thinking, it would be useful to integrate a short survey into the overall exercise. The survey could be run as a final activity with the objective of gaining insight into how CTC affected their understandings of specific Indonesian development issues, as well as how this type of virtual learning exercise altered broader geographical perspectives and outlooks.

Such a survey would need to target both OBU and UI students with a view that any kind of CTC initiative should promote positive learning outcomes for both groups of students. This is to avoid neglecting the benefits for partner groups and students from the Global South which has been observed in the context of school linking initiatives (Leonard, 2005). As such, a reflective exercise of this nature would also help to reinforce the key learning outcomes of CTC and the wider purposes it serves in the development of the learning of all students involved.

Acknowledgments

I would like to thank Paul Lin for his time, energy and enthusiasm in support of the CTC project and, importantly, his permission to use the Oxford Educational Cloud to host the content and student interactions. A sincere thank you to Shofwan Albanna for coordinating the Indonesian end of the project, and to Adam Tyson and Helena Varkkey for their input on the group research tasks. Thank you also to Helen Walkington for her insights on how to turn a fledgling classroom-based idea into something with more meaning and pedagogic value. Finally, thank you to both the Indonesian and UK students for their participation in the project.

Note

1 www.oecglobal.org

References

Clarke, T. and Clarke, E., 2009. Born digital? Pedagogy and computer-assisted learning. *Education + Training*, 51(5/6), pp.395–407.
De Miguel González, R. and Donert, K., 2013. *Innovative Learning Geography in Europe: New Challenges for the 21st Century*. Newcastle-Upon-Tyne: Cambridge Scholars Publishing.
Felicia, P., 2011. *Handbook of Research on Improving Learning and Motivation through Educational Games: Multidisciplinary Approaches*. Hershey, PA: IGI Global.
Ferguson, R., Barzilai, S., Ben-Zvi, D., Chinn, C.A., Herodotou, C., Hod, Y., Kali, Y., Kukulska-Hulme, A., Kupermintz, H., McAndrew, P., Rienties, B., Sagy, O., Scanlon, E., Sharples, M., Weller, M. and Whitelock, D., 2017. *Innovating Pedagogy 2017: Open University Innovation Report 6*. Milton Keynes: The Open University.
Ferguson, R., Coughlan, T., Egelandsdal, K., Gaved, M., Herodotou, C., Hillaire, G., Jones, D., Jowers, I., Kukulska-Hulme, A., McAndrew, P., Misiejuk, K., Ness, I.J., Rienties, B., Scanlon, E., Sharples, M., Wasson, B., Weller, M. and Whitelock, D., 2019. *Innovating Pedagogy 2019: Open University Innovation Report 7*. Milton Keynes: The Open University.
Ferris, S.P. and Wilder, H., 2017. *Unplugging the Classroom: Teaching with Technologies to Promote Students' Lifelong Learning*. Kidlington: Chandos Publishing.
Haigh, M., 2002. Internationalisation of the curriculum: Designing inclusive education for a small world. *Journal of Geography in Higher Education*, 26 (1) pp.49–66.
Kitano, M., 1997. What a course will look like after multicultural change. In A.I. Morey and M. Kitano (eds), *Multicultural Course Transformation in Higher Education: A Broader Truth*. Boston, MA: Allyn and Bacon, pp.18–30.
Leask, B., 2001. Bridging the gap: Internationalizing university curricula. *Journal of Studies in International Education*, 5 (2), pp.100–115.
Leonard, A., 2005. Lessons from UK secondary schools: School linking and teaching and learning in global citizenship and geography. *The Development Education Journal*, 11 (2), pp.36–37.

McEwan, L., 1996. Fieldwork in the undergraduate geography programme: Challenges and changes. *Journal of Geography in Higher Education*, 20, pp.379–391.

O'Hara, M., 2007. Strangers in a strange land: Knowing, learning and education for the global knowledge society. *Futures*, 39, pp.930–941.

Pike, N., 2015. The experimental sciences. In H. Fry, S. Ketteridge and S. Marshall (eds), *A Handbook for Teaching and Learning in Higher Education* (4th edn). London: Routledge.

Potter, R., Binns, T., Elliot, J., Nel, E. and Smith, D., 2018. *Geographies of Development: An Introduction to Development Studies*. London: Routledge.

Chapter 8

Social media as a tool for geographers and geography educators

Francesca Fearnley

Introduction

In March 1989, an engineer and computer scientist at CERN, the European Organisation for Nuclear Research, submitted a proposal for a new IT management system that would link the information stored on all CERN's computers, laying the foundations for the primary tool used to interact on the internet, the World Wide Web (InfoCERN, n.d.). Fast forward 30 years, the internet is now used by an unprecedented number of people, hitting the 4 billion mark in 2018 (Kemp, 2018), and in unprecedented ways including as '… a public square, library, a doctor's office, a shop, a school, a design studio, an office, a cinema, a bank, and so much more' (Berners-Lee, 2019). Over these thirty years, the internet has been on a remarkable journey, with the term Web 2.0 coined in 2004 to describe the new way in which it was being used. Individuals were no longer just consuming content on the internet, but rather helping to create and develop 'user-generated' content in a more collaborative way (Cormode and Krishnamurthy, 2008). With their foundations intertwined with the growth of communications platforms came the invention of social networking sites, such as Facebook, Twitter and Instagram. Of the 4 billion people now accessing the internet, over 3 billion are active users of social media (Kemp, 2018), although it should be noted that the geographical distribution of users is uneven.

But what is the definition of 'social media', how does this differ from 'social networking' and what exactly do these platforms offer? Noting a distinct change in terminology, there was a shift from using 'social networking' in the early 2000s when sites such as Facebook and Twitter were formed, to the use of the term 'social media' in the late 2000s with the growth of other platforms such as Instagram, Pinterest and Snapchat. It is largely accepted that 'social media' is the catch-all term which encompasses these platforms collectively. These are platforms which ultimately allow individuals to 'communicate and collaborate and to create, modify, and share' user-generated content (McCay-Peet and Quan-Hasse, 2017, p.16), which covers everything from video conferencing tools such as Skype to image hosting services like Flickr and networking sites such as LinkedIn.

The following chapter will examine how social media sites, and Twitter in particular, can be used as a data source not only for those working in academic research, but by teachers and students in classrooms and lecture theatres around the world. In doing so, it goes some way towards exemplifying how this modern approach to data collection and analysis might shape the nature of the geographical questions that can be posed by the next generation of geographers, and through this create new opportunities for the 'power of thinking geographically' (Jackson, 2006, p.203) to be capitalised upon within geography education.

Citizens as sensors

Users of social media have been given an identity as information contributors, disseminators and exchangers through the act of being able to create, modify and share content with a wide audience (Crooks et al., 2013). The idea that citizens of the world could act as sensors was first identified by Goodchild (2007) and his theory of 'citizen science'. Goodchild coined the term 'volunteered geographic information' (VGI) to describe the way in which individuals could carry out their day-to-day lives whilst pro-actively collecting data and furthering developments in knowledge, through their detection, reporting, reviewing and responding to of life's events.

Since the conception of Goodchild's (2007) VGI theory, there have been significant technological developments. It is now estimated that two-thirds of the world's population have access to a mobile phone. More than half of these are considered 'smart phones', meaning they support communication and navigational applications, such as the GPS function (Kemp, 2018). Technological developments have been a fundamental driver in the growth of these types of data as smartphones are now the preferred choice of device when going online (rather than desktop computers). Consequently, Goodchild's (2007) theories have been developed further from VGI to 'social media geographic information' (SMGI), a form of crowdsourced geospatial information from individuals who participate in web-based social networking sites (Campagna, 2016). Individuals no longer have to actively 'volunteer' their data (Goodchild, 2007), but produce these data traces 'unconsciously' as part of their everyday internet usage (Capineri, 2016). Content produced online is frequently associated with corresponding geospatial information by the process of geotagging, which assigns a location to the content, commonly in the form of coordinates; geotagging is possible on a variety of sites including Twitter, Facebook and Instagram. Now we will turn to the role of social media within academic research.

Social media, Twitter and research

Using social media for educational purposes has been well considered. From its use as a communication, organisation or information dissemination tool to its use in community building or for assessments, its adoption in educational settings is wide and diverse (see Tang and Hew, 2017, for a comprehensive overview of the many uses of social media in a classroom context). However, it is its purpose as a data collection tool which has gained unprecedented popularity amongst researchers, predominantly in Higher Education.

Research covers a plethora of different platforms to analyse an array of incredibly rich and varied geographical concepts. However, one particular platform, Twitter, a microblogging site, has been heralded by citizens and scholars alike as one of the most popular forms of social media. Twitter works predominantly as a broadcasting function, where individuals post short status updates to their followers, known as tweets. These are message updates no longer than 280 characters in length. This strategy of forcing brevity has meant users post shorter but more frequent mobile updates (Sadilek et al., 2012) which has led to its ability to break news more easily, sometimes before other official sources (Stojanovski et al., 2016), and report images and written content in response to events across the world in near real-time.

When posts are shared on social media, they contain not only the text of the status update, but a range of further associated attributes. In the case of Twitter, attributes include details such as the time the message was created and information about the tweeter, including name and their number of followers. Other attributes include how that information was uploaded (computer or mobile device), whether that tweet was a comment on, reply to, or retweet of another tweet and occasionally coordinates of where the tweet was sent. Research using social media covers at least one, if not more, of these attributes, ranging from time-series investigations to contextual and network analysis. Combining this information with the speed of responses and potential global reach that it offers, Twitter's frequent use in social media related geographic research is unsurprising.

Examining the spatial extent of phenomena

It was the geographical reference associated with tweets that tended to pique geography researchers' interests. Utilising the geotagging functionality to identify a tweet within time and space has allowed them to reveal the characteristics of certain events and phenomena, providing new insights into our world. Disaster and crisis management is one key area that has been utilising social media feeds for this exact purpose. In such time-critical situations, social media networks offer a free and often instantly accessible source of information. A lack of relevant geospatial information has long been recognised as one of the major challenges in emergency management. This was highlighted in the case of Hurricane Katrina, where the White House acknowledged that one of the key failures in their response was a lack of effective communications to properly inform the response and recovery effort (Li and Goodchild, 2010). Consequently, investigations have covered the use of Twitter as a detection and tracking system, in crisis communication, and in investigations of the spatial and temporal extent of natural hazards and other geographic phenomenon including earthquakes (Earle et al., 2011; Crooks et al., 2013), wildfires (Sutton et al., 2008; Kent and Capello, 2013) extreme flooding (Bruns et al., 2012) and global mobility patterns (Hawelka et al., 2014).

Developing situational awareness

A common thread that many articles share is their focus on using Twitter to gain first-hand insight from those on the ground. Earle et al. (2011) noted that the general public very often and frequently within tens of seconds of feeling an earthquake turned to Twitter to share their experiences and to check whether others had felt shaking too. Based on this idea, they used Twitter as an earthquake detection method, detecting 48 earthquakes over the course of five months. The study suggests this method has limitations. For example, 48 earthquakes was a dramatically lower number than the 5000+ detected in the same time frame by the US Geological Survey (USGS). In addition, Earle et al. (2011) could not accurately provide data relating to the location or magnitude of the seismic activity. However, the detection speeds were unprecedented in poorly instrumented regions of the world and they also found that Twitter offered first-hand narratives of on-the-ground experiences, providing a new level of context in rapid time (Earle et al., 2011). This finding echoes that of Lampos and Cristianini (2010), who investigated whether Twitter could be used to track the flu pandemic. They noted that the benefit of using Twitter was the situational awareness that could be gained from tweets, often within a few hours. This was compared to data released by the Health Protection Agency, which often had a delay in releasing data of one to two weeks.

Going beyond the geotag

In 2013, Crampton et al. proposed an extension to the classic practice of mapping SMGI into a new level of investigation, which they termed going 'beyond the geotag' (Crampton et al., 2013, p.130). Rather than producing simple visualisations of the data, placing it within a spatial and temporal context, they called for researchers to situate the investigation within broader geo and sociospatial theory and to understand how it applies in this digital century (Crampton et al., 2013).

Tobler's First Law of Geography established that 'everything is related to everything else, but near things are more related than distant things' (Tobler, 1970, p.236), a theory similar to that of the classic distance decay function (cultural or spatial interaction decreases as distance increases). Over 30 years later in 2005, Thomas Friedman published the book *The World Is Flat*, remarking to *Wired* Magazine the same year that 'Several technological ... forces have converged, and that has produced a global, Web-enabled playing field that allows for multiple forms of collaboration without regard to geography or distance' (Pink, 2005). Such work had similar themes to that of Cairncross (2001), who had argued that advances in communications technology had rendered distance 'dead' (at least for those with access to such technology). Both of these theories proposed that interactions within society were no longer bounded by geographic proximity. Demonstrating how going 'beyond the geotag' (Crampton et al., 2013, p.130) could be achieved, in their investigation of the geography of tweets sent during rioting in Kentucky in 2011, they found that the classic distance decay function applied to the spatial diffusion of information but within a refined local context and not much further. The study appeared to suggest that Tobler's Law still applied in a digital era. On the other hand, Bruns et al. (2012) suggested that emotionally charged events cause a reaction in digital space that transcends local boundaries. For example, Crawford (2009) investigated the spread of users turning their Twitter icons green in an expression of solidarity for the protest movement in Iran. The researchers found that the response diffused far beyond Iran, with a significant uptake amongst international Twitter users (Crawford, 2009). Similarly, Shelton et al. (2014) found that swathes of tweets sent in response to Hurricane Sandy were concentrated in areas most affected by the storm (in the Northeastern US). However, they also saw the information spread past this immediate local context with a distinct cluster of intense tweeting around Phoenix, Arizona. Such research provides an opportunity to delve deeper to understand relational sociospatial networks.

It is investigations which question and debate these ideas and go beyond mapping towards understanding how their work can contribute to and update geographical thinking which challenge the way that geographical knowledge is produced, diffused and understood in this modern age (Capineri, 2016). Despite their wide variations in theme, the articles above demonstrate common ground in their methods and, ultimately, their very approaches to asking questions and presenting results, which are lessons that are widely applicable to a geography classroom.

Limitations of social media research

Like all methods of data collection, research using social media has distinct and often unique limitations which need to be taken into consideration by researchers at all levels. The following section provides a broad understanding of a few common limitations, when using

Twitter in particular, but is by no means a complete or exhaustive list of the limitations of all social media sites.

Representation

We know that at the end of 2018, Twitter had 321 million monthly active users (Twitter, 2018a), but what do we know about who these users are? Multiple studies have demonstrated that far from being a true representation of society, Twitter users are typically of a wealthy, educated, white, Western and male demographic (Haklay, 2012). SMGI in general tends to be sent from more urban areas (Capineri, 2016). Crutcher and Zook (2009, cited in Shelton et al., 2014, p.169) highlight that SMGI was produced more frequently in response to Hurricane Katrina in 'wealthier, whiter, more tourist-oriented locations within New Orleans despite the greatest effects of the storm being felt in predominantly poor and black areas'. There are significant practical implications that can be drawn from this. If disaster response officials conclude that a greater concentration of tweets in one area means the area has had a higher level of damage and therefore needs more disaster relief, then the social inequalities that are perpetuated in physical space are reinforced. Rather, harder hit and more damaged areas may have lost access to technology such as WiFi or data connectivity on mobile devices, which causes a lull in the creation of content (Shelton et al., 2014). Such ideas demonstrate the need for researchers to have a comprehensive understanding of uneven technological access and use in order to ensure that their conclusions do not reinforce inequality through misrepresenting people and places.

Research has also shown that 2.9 billion people access social media via mobile devices out of around 3 billion active social media users. However, access across the world remains unevenly distributed, with some populations dramatically underrepresented. This is true across regions such as Eastern Africa and Southern Asia, with only 27% and 36% of their populations respectively using the internet (Kemp, 2018). Again, researchers need to bear these factors in mind and ensure they do not attempt to use the conclusions drawn from their research to make assumptions about the population as a whole.

Further sampling

In addition to limitations around representation in terms of active Twitter users, not all tweets generated are made available to the public, meaning that the data made available through Twitter's application programming interfaces (APIs) is almost certainly a subset of the original. How this is determined and on what basis is unknown, but for its standard (free) API packages, it is suggested that it could be a 1% sample of all tweets (Crooks et al., 2013). Twitter (n.d.) offers 'scalable access to increased data' in its Premium package, and 'the highest level of access' via its Enterprise package. Based on a tiered system depending on the scale of your request, access to the top tier could cost around $2500 per month (Perez, 2017).

Furthermore, whether an individual user uses a desktop or mobile device to access Twitter also has an impact on the total possible number of individuals whose geolocations should, in theory, be accurate. There are two ways in which someone could georeference their tweet, the first being a self-defined location in the user's biography. However, there are not any regulations in place to ensure this has to be a real location (so an individual could use 'Middle-Earth' or 'The Moon' as their location). Those that are real locations are often defined at city or country level. The other way is the precise coordinates geotagged to the tweet itself, derived from GPS. This method is more accurate; however, users have to 'opt-in' to allow this functionality. Uptake is understandably low, which means that only between 1 and 1.7% of tweets contain precise associated coordinates (Shelton et al., 2014; Stojanovski et al., 2016). It is this level of iterative sampling which means that researchers have to take care with the conclusions that they draw from their investigations.

Data privacy

Consent is central in ethical reviews of research involving human participants. In the majority of cases, this takes the form of informed consent which involves providing a participant with details of what the study will be about, how the participant will be involved and to make them aware that should they wish to, they can withdraw at any time. However, when data sets involve the data of potentially millions of users, generally made publicly available on the internet, how do these ideas apply?

Rather than offering customisable privacy settings like other social media sites such as Facebook, Twitter has a simple 'public' or 'private' choice which covers the entirety of a user's profile. By default, once an account has been created, one's profile and any content generated is made public and only through the manual changing of settings can the account be made private. Consequently, it is estimated that fewer than 10% of Twitter accounts are private (Zimmer and Proferes, 2014). When an individual posts content through Twitter, they grant Twitter the 'worldwide, non-exclusive, royalty-free license … to use, copy, reproduce, process, adapt, modify, publish, transmit, display and distribute such Content' (Twitter, 2018b). Such clauses in Twitter's Terms of Service allow researchers to access a user's data, a process that is governed by strict policies. Subsequently, there has been debate amongst researchers about whether they should be able to access the data provided through public accounts and to what extent those with public accounts are aware that their data may be harvested and analysed for research.

Some consider data made available online as being 'in the public domain', suggesting that consequently researchers do not need to ask permission to harvest this data, as by creating the content users agree to it (see, for example, Manovich, 2011). Others such as Zimmer (2010) argue that 'there is a reasonable expectation that one's tweet stream will be "practically obscure" within thousands (if not millions) of tweets', and therefore users do not automatically consent to their data being harvested and used by researchers. Beninger (2017) explored the views social media users have on the matter, demonstrating they were wide, diverse and dependent on a variety of factors including the purpose of the research, the affiliation of the researcher (public institution vs private corporation) and whether their anonymity would be ensured, which is a conclusion that 'muddies the water, making the application of traditional ethical principles more challenging' (Beninger, 2017, p.71).

Such debates are hard to navigate even for experienced academics. Consequently, the research ethics committees of institutions around the world are developing best practice guidance for internet-based research. These provide localised advice which balances the potential benefits of gaining new scholarly insight with protecting the privacy and rights of the users (McCay-Peet and Quan-Hasse, 2017). However, the creation of guidelines should not just be applicable to academic professionals. To ensure clarity and transparency for students, it is the responsibility of those in work-setting roles at all levels (such as teachers and dissertation supervisors) to develop their own in-house guidance that sets out how each school or department approaches issues such as data privacy. In order to satisfy the requirements of such committees, solutions proposed include anonymising results and focusing data collection at an aggregate level, rather than on an individual level. Considering that scale forms a key component of geographical thinking, an appreciation of the need for sensitivity when handling such data is likely to shape the nature of questions that can be asked before doing such research; indeed, an 'awareness of ethical questions' (Lambert and Reiss, 2016, p.31), in and of itself, is important within fieldwork in school geography. These issues will undoubtedly have an impact on the types of enquiries students will be able to pursue.

GDPR

In addition to the rules and policies which govern Twitter, these debates now sit against a backdrop of the introduction of the EU General Data Protection Regulations (GDPR). Introduced to protect and empower the data privacy of all EU citizens, GDPR allows individuals to better control the way companies and organisations use their data. Those seen to be in breach of the regulations can now be fined anywhere up to 4% of their annual global turnover or 20 million Euros, whichever is greater (EU GDPR, n.d.). There has been no data breach more significant in the 21st century which has highlighted the need for tightening of such regulation than that of the Facebook/Cambridge Analytica scandal, where the data of 50 million Facebook users was unknowingly harvested and allegedly used to influence voting in key elections including the UK Brexit Referendum and 2016 US Presidential Election. Such controversies have ultimately led Mark Zuckerberg to declare his vision for social networking as one that is privacy-focused, declaring this future for the internet as more important than the open platforms of today. Likely to take the form of a simpler, private platform, major changes could include end-to-end encryption which would prevent even the developers from being able to access what is being shared (Zuckerberg, 2019). Such changes could signal the start of a new era shaped by changes in the way researchers are able to access and analyse such data streams, something that teachers, lecturers and those in educating roles should pay close attention to.

SMGI as a tool of research for geographers and geography educators

A defining feature that allows geographers to stand out amongst those from other subject areas is a unique understanding of concepts such as space, place, time and scale (Clifford et al., 2008). By developing SMGI as a tool of research within the discipline (compared to other classic disciplines that study social media such as communications, IT or mathematics), an opportunity arises to address new questions and provide new perspectives.

The applications are wide and varied, from empowering students to carry out primary data collection at GCSE and A-Level (especially relevant with more recent specifications that require non-examined independent investigations) through to undergraduate dissertations, or to enrich class-based discussions and teaching. Such an opportunity empowers teachers, and more so young adults, to stretch themselves by learning new methodologies, identifying areas of the discipline that resonate with them and shaping themselves as researchers along the way. The value that SMGI brings is not just in the conclusions that can be drawn from such investigations and the subsequent contribution to geographical knowledge, but rather in the entire process, from question inception through to data analysis and research evaluation (in terms of limitations and accuracy).

Take for example the data collection aspect. There are many ways in which this can be done, ranging from simply cutting and pasting from the website itself to purchasing the data directly from Twitter. One of the most commonly utilised options involves using basic coding skills and programming languages such as R and Python to access the website's APIs, where such data can be downloaded. There are many websites and video tutorials that can assist with this process. Coding skills have long been identified as a key method of encouraging a problem-solving, solutions-driven mindset amongst members of a society (Tuomi et al., 2018). Initiatives in the UK such as the BBC's 'Make it Digital' campaign have acknowledged the need to help build digital skills within specific groups as 'the best way to help them thrive' (BBC, n.d.). Why not use opportunities within geographical education as a vehicle to do the same, in turn opening up opportunities for interdisciplinary collaboration?

Analysing this data also requires the introduction of new quantitative and qualitative methods, some of which are not frequently associated with the discipline. From quantifying simple metrics, such as SMGI plotted over time and calculating the unique number of users, to data composition analysis (which could identify components such as the number of original tweets, retweets or conversations in a particular subset of information), these metrics provide researchers with fascinating baseline results, alongside classic techniques such as mapping. Going beyond the 'mapping' of phenomena within a spatial and temporal context, analysis methods include techniques such as sentiment and time-series analysis, facilitated by the use of tools such as SentiStrength (an algorithm which estimates the strength and direction of sentiment). Relational network analysis can help provide further in-depth results and be used to support and update geographical thinking. Content analysis techniques such as word cloud generators, which identify key words and phrases and plot their size based on frequency of use, can help to provide qualitative context. Word cloud generators have been suggested as a way to analyse the words that people choose to use in their social media posts in association with certain places in order to understand more about how a place is represented (Oakes, 2018). Such analytical methods can be done manually, or again through programming packages such as the word cloud generation package in R.

Furthermore, there is great value in being able to carry out desk-based research to help illustrate geographical phenomena in a teaching context. For example, a teacher in Norfolk is discussing natural hazards when an unprecedented flash flood event occurs in Western Australia. Using her skills in social media research, she is able to create a database of tweets that have been sent in the lead up to, during and after the flash flooding event. Working collaboratively with her students, she is able to paint a picture to her classroom of the severity of the floods, first-hand accounts of how they happened, the known impact so far and the quality of the response – possibly before such accounts are being broadcast by official news sources. A few months later when it comes to designing his own research investigation, one student thinks back to when his teacher used this exciting new data collection method and has the confidence to attempt something similar in his own work. A new tourist attraction has opened in his hometown. He wonders whether he could use the same techniques to explore the impact this attraction is having and the perceptions of local residents. Mapping the tweets, he is surprised by a spike in responses from a small town in Germany, an opportunity which enables him to 'go beyond the geotag' and investigate the possibility of a new sociospatial relationship.

Whether in class, or by encouraging students of all levels to explore SMGI research as a plausible and effective option, such research exposes individuals in the discipline of geography to a host of new skills, including research and analysis methods which will broaden and deepen their understanding of what geography is and what it encompasses. All applications have the power to extend learning opportunities beyond the classroom and into a real-world setting (Tang and Hew, 2017). While the current state of the education system imposes constraints of time and money, factors which ultimately limit opportunities for field-based study, the opportunity should be taken to enrich student learning with this free, easily accessible and rich data source.

Conclusion

Social media, and Twitter in particular, offers an easily accessible and rich source of geographical information and commentary. The limitations and ethical considerations of using social media data are likely to shape the nature of future investigations. At present, the ability to gain contextual information and associated commentary in an instant from anywhere in the world is unprecedented, which means that a class of students at the heart of the English countryside have the ability to monitor the reactions of an event in real time on an entirely different continent, bringing us all closer to the action and putting us in a better position to understand place/people interactions than ever before.

Far from being a research tool that can only be used by experts, there is no doubt that the approach to asking questions, collecting data and analysing results presented here could, and should, be applied by geographers and geography teachers at all levels of education. After all, for both teachers and students alike, the process provides an opportunity to learn new skills, develop a repertoire of methods for geographical research, and ask questions that enhance their capacity to 'think geographically'.

References

Beninger, K., 2017. Social media users' views on the ethics of social media research. In L. Sloan and A. Quan-Hasse (eds), *The Sage Handbook of Social Media Research Methods*. London: Sage, pp.57–73.

Berners-Lee, T., 2019. 30 years on, what's next #ForTheWeb? *Webfoundation.org News and Blogs*. Available at: https://webfoundation.org/2019/03/web-birthday-30/ [Accessed 14 March 2019].

British Broadcasting Corporation (BBC), n.d. *Make It Digital: How the BBC and its Partners Unleashed the UK's Digital Creativity*. Available at: www.bbc.co.uk/programmes/articles/1Q5xGB0YnVf2CgTPyL0MK2s/make-it-digital-how-the-bbc-and-its-partners-unleashed-the-uks-digital-creativity [Accessed 16 April 2019].

Bruns, A., Burgess, J.E., Crawford, K. and Shaw, F., 2012. *#Qldfloods and @QPSMedia: Crisis Communication on Twitter in the 2011 South East Queensland Floods*. Brisbane: ARC Centre of Excellence for Creative Industries and Innovation. Available at: http://eprints.qut.edu.au/48241/1/floodsreport.pdf [Accessed 12 February 2019].

Cairncross, F., 2001. *The Death of Distance: How the Communications Revolution is Changing Our Lives*. Boston, MA: Harvard Business School Press.

Campagna, M., 2016. Social media geographic information: Why social is special when it goes spatial? In C. Capineri, M. Haklay, H. Huang, V. Antoniou, J. Kettunen, F. Ostermann and R. Purves (eds), *European Handbook of Crowdsourced Geographic Information*. London: Ubiquity Press, Chapter 4.

Capineri, C., 2016. The nature of volunteered geographic information. In C. Capineri, M. Haklay, H. Huang, V. Antoniou, J. Kettunen, F. Ostermann and R. Purves (eds), *European Handbook of Crowdsourced Geographic Information*. London: Ubiquity Press, Chapter 2.

Clifford, N., Holloway, S., Rice, S.P. and Valentine, G. (eds), 2008. *Key Concepts in Geography*. London: Sage.

Cormode, G. and Krishnamurthy, B., 2008. Key differences between Web 1.0 and Web 2.0. *First Monday*. Available at: https://firstmonday.org/ojs/index.php/fm/article/view/2125/1972 [Accessed 14 March 2019].

Crampton, J.W., Graham, M., Poorthuis, A., Shelton, T., Stephens, M., Wilson, M.W. and Zook, M., 2013. Beyond the geotag: Situating 'big data' and leveraging the potential of the geoweb. *Cartography and Geographic Information Science*, 40 (2), pp.130–139.

Crawford, K., 2009. *Everything's Gone Green*. Available at: www.abc.net.au/news/2009-06-24/28486 [Accessed 15 March 2019].

Crooks, A., Croitoru, A., Stefanidis, A. and Radzikowski, J., 2013. #Earthquake: Twitter as a distributed sensor system. *Transactions in GIS*, 17 (1), pp.124–147.

Crutcher, M. and Zook, M., 2009. Placemarks and waterlines: Racialized cyberscapes in post-Katrina Google Earth. *Geoforum*, 40 (4), pp.523–534.

Earle, P.S., Bowden, D.C. and Guy, M., 2011. Twitter earthquake detection: Earthquake monitoring in a social world. *Annals of Geophysics*, 54 (6), pp.708–715.

EU GDPR, n.d. *GDPR Key Changes*. Available at: https://eugdpr.org/the-regulation/ [Accessed 19 April 2019].

Goodchild, M.F., 2007. Citizens as sensors: The world of volunteered geography. *GeoJournal*, 69 (4), pp.211–221.

Haklay, M., 2012. 'Nobody wants to do council estates': Digital divide, spatial justice and outliers – AAG 2012. *Po Ve Sham – Muki Haklay's Personal Blog*. Available at: https://povesham.wordpress.com/2012/03/05/nobody-wants-to-do-council-estates-digital-divide-spatial-justice-and-outliers-aag-2012/ [Accessed 27 February 2019].

Hawelka, B., Sitkoa, I., Beinat, E., Sobolevsky, S., Kazakopoulos, P. and Ratti, C., 2014. Geo-located Twitter as proxy for global mobility patterns. *Cartography and Geographic Information Science*, 41 (3), pp.260–271.

InfoCERN, n.d. *Tim Berners-Lee's Proposal*. Available at: http://info.cern.ch/Proposal.html [Accessed 12 February 2019].

Jackson, P., 2006. Thinking geographically. *Geography*, 91 (3), pp.199–204.

Kemp, S., 2018. Digital in 2018: World's internet users pass the 4 billion mark. *WeareSocial.com Blog*, 30 January. Available at: https://wearesocial.com/blog/2018/01/global-digital-report-2018 [Accessed 28 March 2019].

Kent, J.D. and Capello, H.T., 2013. Spatial patterns and demographic indicators of effective social media content during the Horsethief Canyon fire of 2012. *Cartography and Geographic Information Science*, 40 (2), pp.78–89.

Lambert, D. and Reiss, M.J., 2016. The place of fieldwork in geography qualifications. *Geography*, 101 (1), pp.28–34.

Lampos, V. and Cristianini, N., 2010. Tracking the flu pandemic by monitoring the social web. In *IEEE 2nd International Workshop on Cognitive Information Processing. Elba Island, Italy 14–16 June 2010*. Washington, DC: IEEE.

Li, L. and Goodchild, M.F., 2010. The role of social networks in emergency management: A research agenda. *International Journal of Information Systems for Crisis Response and Management*, 2 (4), pp.49–59.

Manovich, L., 2011. *Trending: The Promises and the Challenges of Big Social Data*. Minneapolis, MS: Manovich. Available at: http://manovich.net/content/04-projects/067-trending-the-promises-and-the-challenges-of-big-social-data/64-article-2011.pdf [Accessed 14 March 2019].

McCay-Peet, L. and Quan-Haase, A., 2017. What is social media and what questions can social media research help us answer? In L. Sloan and A. Quan-Haase (eds), *The Sage Handbook of Social Media Research Methods*. London: Sage, pp.13–26.

Oakes, S., 2018. *Fieldwork at A-Level: Your Guide to the Independent Investigation Guide*. Sheffield: Geographical Association.

Perez, S., 2017. *Twitter Launches Lower-Cost Subscription Access to its Data through New Premium APIs*. Available at: https://techcrunch.com/2017/11/14/twitter-launches-lower-cost-subscription-access-to-its-data-through-new-premium-apis/ [Accessed 5 March 2019].

Pink, D.H., 2005. *Why the World is Flat*. Available at: www.wired.com/2005/05/friedman-2/ [Accessed 14 March 2019].

Sadilek, A., Kautz, H. and Silenzio, V., 2012. Modelling spread of disease from social interactions. In *AAAI, Proceedings of the Sixth International AAAI Conference on Weblogs and Social Media. Dublin, Ireland 4–7 June 2012*. Palo Alto, CA: The AAAI Press.

Shelton, T., Poorthius, A., Graham, M. and Zook, M., 2014. Mapping the data shadows of Hurricane Sandy: Uncovering the sociospatial dimensions of 'big data'. *Geoforum*, 52, pp.167–179.

Stojanovski, D., Chorbev, I., Dimitrovski, I. and Madjarov, G., 2016. Social networks VGI: Twitter sentiment analysis of social hotspots. In C. Capineri, M. Haklay, H. Huang, V. Antoniou, J. Kettunen, F. Ostermann and R. Purves (eds), *European Handbook of Crowdsourced Geographic Information*. London: Ubiquity Press, Chapter 17.

Sutton, J., Palen, L. and Shklovski, I., 2008. Backchannels on the front lines: Emergent uses of social media in the 2007 southern California wildfires. In *ISCRAM, Proceedings of the Fifth International Conference on Information Systems for Crisis Response and Management. Washington DC, 4–7 May 2008*. Washington, DC: ISCRAM.

Tang, Y. and Hew, K.F., 2017. Using Twitter for education: Beneficial or simply a waste of time? *Computers and Education*, 106, pp.97–118.

Tobler, W.R., 1970. A computer movie simulating urban growth in the Detroit region. *Economic Geography*, 46, pp.234–240.

Tuomi, P., Multisilta, J., Saarikoski, P. and Suominen, J., 2018. Coding skills as a success factor for a society. *Education and Information Technologies*, 23 (1), pp.419–434.

Twitter, n.d. *API Access that Scales with You and Your Solution*. Available at: https://developer.twitter.com/en/pricing [Accessed 29 December 2019].

Twitter, 2018a. *Selected Company Metrics and Financials*. Available at: https://s22.q4cdn.com/826641620/files/doc_financials/2018/q4/Q4-2018-Selected-Company-Financials-and-Metrics.pdf [Accessed 12 March 2019].

Twitter, 2018b. *Twitter Terms of Service*. Available at: https://twitter.com/en/tos#intlTerms [Accessed 15 April 2019].

Zimmer, M., 2010. *Is it Ethical to Harvest Public Twitter Accounts without Consent?* Available at: www.michaelzimmer.org/2010/02/12/is-it-ethical-to-harvest-public–twitter-accounts-without-consent/ [Accessed 26 February 2019].

Zimmer, M. and Proferes, N., 2014. Privacy on Twitter, Twitter on privacy. In K. Weller, A. Bruns, J. Burgess, M. Mahrt and C. Puschmann (eds), *Twitter and Society*. Oxford: Peter Lang, pp.169–182.

Zuckerberg, M., 2019. *A Privacy-Focused Vision for Social Networking*. Available at: www.facebook.com/notes/mark-zuckerberg/a-privacy-focused-vision-for-social-networking/10156700570096634/ [Accessed 27 April 2019].

Part III

Geospatial technologies in the digital world

Chapter 9

Insights from professional discourse on GIS

A case for recognising geography teachers' repertoire of experience

Grace Healy

Introduction

Whilst the place of GIS is well established within geography curricula in England (DfE, 2013), there is ongoing concern over how both beginning and experienced teachers can develop their practice with GIS (GA, 2011; Walshe, 2017; Collins and Mitchell, 2019). With this in mind, a focus upon professional discourse opens up scope to explore the repertoire of experience with GIS found within the geography education community. Persaud (2011) emphasises that professional journals provide 'powerful "discourses" – ways of speaking – which can help readers to frame and put into perspective their own practices' (p.137). For the purpose of this study, professional journals are utilised as a way to illuminate how GIS appears to be negotiated within the geography education community, acknowledging that they provide a way for teachers both to use and contribute to knowledge (Ellis, 2007).

This chapter draws upon a review of how GIS is portrayed within the discourse of *Primary Geography* and *Teaching Geography*, both flagship journals of the Geographical Association (GA) aimed at providing support and guidance for early years and primary teachers, and secondary and college teachers respectively. Whilst the GA states on its website that it is an 'international community of practice for all geography teachers', it is a UK-based organisation and the majority of published articles within these professional journals relate to the UK context. Therefore, this review focuses more explicitly on professional discourse within the geography education community in the UK. The third journal of the GA (*Geography*) was not included within this review because a preliminary search of articles found that the authorship of articles relating to GIS was predominantly from academic geographers. Whilst this would have added a different dimension to this review, it was decided that it would not directly contribute to an understanding of how GIS was being negotiated by geography teachers within the geography education community.

In this chapter, professional discourse is positioned as a way of understanding geography teachers' repertoire of experience with GIS. It is proposed that professional discourse provides significant insights into GIS as an aspect of geography education in the digital world, because of the way in which geography teachers' beliefs, values and attitudes influence their use of GIS (Höhnle et al., 2016; Hong and Stonier, 2015; Walshe, 2017) and the immediacy with which geography teachers now have access to advancing GIS technologies (Ricker and Thatcher, 2017). A case is made for recognising the collective role of practitioners and researchers in developing geography teachers' professional practice with GIS.

GIS in initial teacher education and continued professional development

Teachers' professional development is noted as a key strand of Baker et al.'s (2015) research agenda for geospatial technologies and learning. For example, Jo (2016) highlights that teachers' dispositions towards GIS are often neglected in GIS research; whilst Walshe (2017) has emphasised the importance of teachers valuing how GIS could enhance their geography teaching. In response to this, it is anticipated that professional discourse that is written by and for members of the geography education community might provide a way to foreground teachers' beliefs, values and dispositions towards GIS.

Within initial teacher education (ITE), Walshe (2017) highlights that trainee teachers can be disempowered by their placement schools' cultures and their mentors' attitudes towards GIS. This resonates with the work of Counsell et al. (2000) and Brooks et al. (2012) which highlights the influential relationship mentors and mentoring departments have on trainee teachers' practice more generally. Walshe (2017) found that trainees appeared to value most the opportunity to see how a practising geography teacher used GIS to support geographical thinking in their geography lessons. Using a vignette, Walshe (2017) exemplifies how one trainee teacher began to recognise the value of GIS for their geography teaching through engaging with how fellow trainee teachers had used GIS successfully within their practice. As highlighted by Persaud (2011) and Counsell (2012), professional discourse provides scope for beginning and experienced teachers to reflect on their own and others' teaching practice. Therefore, it is worthy of consideration to establish whether professional discourse appears to enable geography teachers to appreciate how GIS can support geographical learning and, therefore, learn from the repertoire of experience held by their subject community.

Höhnle et al. (2016) developed a comprehensive overview of the features of effective continued professional development (CPD) in GIS for teachers, drawing upon a survey of constraints acknowledged by teachers. The ten features of effective CPD were introduced and then validated against the recommendations for GIS CPD within five earlier studies (e.g. Kidman and Palmer, 2006). The development of professional learning communities was proposed as one feature; for example Höhnle et al. (2016) suggested at least three teachers from every school should be involved, and there should be an opportunity to consider the findings of classroom research, so as to motivate teachers by illuminating the value of using GIS for students. This recommendation also addresses Jo's (2016) and Walshe's (2017) concerns, as this provides a mechanism by which teachers can recognise how GIS can enhance student learning in geography (Höhnle et al., 2016).

Despite incorporating the structural features of effective GIS by Höhnle et al. (2016), Collins and Mitchell (2019) found there was still a disconnect between this professional development and long-term, sustained and embedded use of GIS within teachers' classroom practice. In response, they posited two significant recommendations (Collins and Mitchell, 2019): the first is the importance of laying the foundations for GIS use within ITE; and the second is a longer-term approach with continuous follow-up and coaching to enable sustained integration of GIS within teachers' classroom practice. I suggest that professional discourse might be one approach to embedding this as it provides an ongoing mechanism for teachers' professional learning and also can illuminate how teachers have engaged with GIS within the context of their own professional practice.

Teachers' technological pedagogical content knowledge and subject knowledge

Teachers' preparedness to capitalise upon GIS within geography education has been framed through the model of technological pedagogical content knowledge (TPACK; Mishra and Koehler, 2006; Koehler and Mishra, 2009). The TPACK model characterises the need for teachers to master this knowledge base, and be attentive to the interaction between technology, pedagogy and content (Mishra and Koehler, 2006), and has been drawn upon by numerous geography educators (e.g. Doering et al., 2009; Hong and Stonier, 2015; Walshe, 2017; Collins and Mitchell, 2019). Oda et al. (2019) have highlighted that there are three distinctive epistemological views of TPACK:

1. TPACK is perceived as an extension of pedagogical content knowledge (PCK; Shulman, 1986) and the technological knowledge base is incorporated separately (Cox and Graham, 2009).
2. TPACK requires integration of each knowledge base and, therefore, it can be integrated into teachers' practice when they use different technologies (Koehler and Mishra, 2009).
3. TPACK is treated as a conglomerate and teachers need time to assimilate each new technology before they can immediately integrate this technology into their classroom practice.

In the context of this study, it is relevant to reflect on whether there is evidence of these epistemological views being held in professional discourse, because of their implications for the pace of integration of GIS within geography classrooms. There has also been less consideration of how the TPACK model works in the context of primary geography education (Gómez Trigueros, 2018). Walshe (2017) proposes that web-based GIS ought to be incorporated within geography curricula from primary education, but also acknowledges the challenges of this given the lack of geography subject specialists within the primary sector, which means there are limits to teachers' geographical knowledge base in relation to both pedagogical and content knowledge.

Within geography education, there has also been a concern around teachers' subject knowledge; Bednarz and Ludwig (1997) and Bednarz and van der Schee (2006) have signalled that teachers' understanding of the discipline of geography needs to be secure before they can fully appreciate how GIS can enhance students' geographical education. More recently, there has been a shift towards focusing upon the extent to which students are supported in understanding how geographical knowledge is constructed in GIS (Fargher, 2018) and how engagement with GIS can develop students' geographical knowledge and thinking (Fargher, 2017; Walshe, 2018; Healy and Walshe, 2020). These shifts further illuminate how an understanding of the discipline of geography interacts with the role of GIS for students' learning in geography. Fargher (2018) argues that, in order to ensure that the potential of GIS is realised, there must be attention to how it contributes to students' powerful geographical knowledge where the 'significance of geographical context and the specificity and interconnectedness of places(s)' (p.24) can be capitalised upon. This has implications for both ITE and CPD, as Fargher (2018) asserts the need for professional development in GIS to be developed in relation to teachers' expertise in geography curriculum making. At the same time there is acknowledgment that GIS education needs to be developed to be responsive to the rapidly evolving GIS landscape, so that geography students (both school and undergraduate) understand principles and concepts in relation to

GIS regardless of the particular functionality of GIS programmes (Ricker and Thatcher, 2017).

In the context of exploring master's level practitioner research, Brooks (2018) argued that practitioners' focus on the 'problems of the day' needed to be acknowledged within the academic field of geography education. The body of practitioner research explored within Brooks' (2018) research illustrates how 'Interests shift. Priorities shift. Insights shift' (Slater, 1995, p.6), and how this can reveal the concerns of geography teachers. With this in mind, this chapter hopes to elicit both whether the themes dominant in academic literature are also apparent in professional discourse within the geography education community, and whether professional discourse can provide insights into how GIS is been navigated by geography teachers. This study will address this through the following research questions:

1 How does professional discourse relate to academic literature around GIS?
2 What is revealed about teachers' professional practice in relation to GIS?

The research approach

Once all articles that had a substantive focus upon GIS across *Primary Geography* since 2005 and *Teaching Geography* since 1975 had been identified, content analysis was carried out – a process of open coding was undertaken through which a set of classification categories emerged from the data (inductive coding analysis). This was reviewed iteratively to increase validity. This research then explored a second set of data; this second stage involved identifying where GIS was mentioned within articles from *Primary Geography* and *Teaching Geography* but was not established as the main focus of the text. The analysis for these articles was informed by van Manen's (1990) level of thematic analysis, which involved looking at the level of words, sentences and the text as a whole; this allowed for the identification of revealing phrases used in connection to GIS. A process of a priori coding was undertaken whereby the classification categories developed in the first stage of the research were applied. There were a number of references to GIS that did not fit with the themes found in the first stage of the research, and therefore additional categories were introduced to capture these representations of GIS within professional discourse. Whilst not employing discourse analysis, it is still necessary to acknowledge that texts from within professional journals do not merely depict reality but also play a role in creating and sustaining it (Denscombe, 2007). In the following discussion when articles from *Primary Geography* and *Teaching Geography* are referred to this is denoted by (PG:) and (TG:) before the in-text citation.

Characteristics of professional discourse on GIS

Since 1975, 19 articles were published in *Teaching Geography* that focused upon GIS (Table 9.1), and a further 37 articles mentioned GIS in relation to other aspects of geographical education. 22 unique authors produced the 19 articles. The professional biography included with each article enables consideration of the stakeholders that are contributing to this professional discourse around GIS. Eight of the articles published were written by at least one practising secondary geography teacher. This includes geography teachers at different stages of their career, including heads of department and senior leaders. Five articles were written by consultants or people working for commercial

Table 9.1 Journal articles from *Primary Geography* (PG) and *Teaching Geography* (TG) with a focus on GIS.

Year	Title
1993 TG	Getting started with GIS
1994 TG	A guide to geographical information systems (GIS)
1995 TG	Implementing GIS technology in schools
2000 TG	Teaching and learning with geographical information systems
2000 TG	Getting start with GIS
2003 TG	GIS in secondary geography
2006 TG	Talk the talk: Mapping mobile phone masts with GIS consultant reflecting on practice
2006 PG	A new look at GIS in the primary classroom
2006 PG	Putting the world in the palm of their hand
2007 TG	Progressive GIS
2008 PG	HOTShot GIS
2009 TG	The application of UK web-based GIS in geography teaching
2010 TG	Why use GIS?
2011 TG	Beyond 2012: Using GIS to investigate the sustainability of the Olympic stadium
2015 TG	A world-wide geographical investigation using online GIS
2016 TG	Developing an explosive scheme of work
2016 TG	Using ArcGIS Online Story Maps
2017 TG	How to ... integrate GIS effectively into the curriculum
2018 TG	Digimap for schools – what is it and what can you do with it?
2018 TG	GIS has changed! Exploring the potential of ArcGIS Online
2018 TG	Glacial landforms: A teaching resource in maps and GIS
2019 TG	Real-world geographers and GIS: Relevance, inspiration and developing geographical knowledge

organisations; a further four articles were written by academic geographers; and two articles were written by university-based teacher educators. Whilst, it has already been established that the journals are bound within the UK context, often these articles were contextualised within the English education system. Three articles were published in *Primary Geography* since 2005 that held a principal focus on GIS and a further four articles mentioned GIS. The three main articles were produced by three unique authors: a teacher, a university-based teacher educator and someone working for a commercial organisation.

Since 2016, articles have shifted from references, to use of a desktop GIS, to WebGIS, such as Digimap and ArcGIS Online, and related applications such as Story Maps. Baker (2015) posits that WebGIS enables geographical learning through the use of GIS rather than merely learning how to use GIS, as the technological burden is reduced. Ricker and Thatcher (2017) acknowledge that the rapidly evolving GIS landscape has affected the development of pedagogy within HE. This evolving GIS landscape might explain why only two articles reference earlier contributions to professional discourse on GIS from *Primary Geography* and *Teaching Geography*. Within the professional discourse of *Primary Geography* and *Teaching Geography*, there is not an expectation that contributions will draw on scholarship from academic discourse. From the articles with a principal focus on GIS within *Primary Geography* and *Teaching Geography*, there is only one that includes an explicit reference to academic discourse on GIS.

Themes of professional discourse on GIS

This section introduces the six categories that emerged as being dominant within professional discourse on GIS. These are not going to be explored in detail here, but will be drawn upon within the subsequent discussion framed by the study's research questions. The themes were:

- Navigating new GIS technologies.
- Reflecting on GIS within geography teaching and curriculum development.
- Reflecting on department learning in relation to GIS.
- GIS in relation to geographical enquiry.
- GIS as a mechanism to develop spatial literacy.
- GIS as a part of geography fieldwork.

These themes were not mutually exclusive, and several articles foregrounded a focus that brought together two or more of these categories. Taking into account articles where GIS was a subsidiary consideration or was mentioned in passing, there were articles where these themes continued to be relevant, but there were also different ways in which GIS was framed that were not represented by these themes, such as references to its importance as a geographical skill and in relation to career opportunities for students. The additional ways in which GIS was been discussed by authors in *Primary Geography* and *Teaching Geography* have been integrated into discussion framed by the second research question below.

How does professional discourse relate to the academic literature around GIS?

Within *Teaching Geography*, the earlier articles focused on introducing new technologies and how they might be used by teachers, exploring questions like 'What is a computer GIS?' (TG: Freeman et al., 1994, p.36; PG: Kelly, 2006) and considering what different GIS skills, such as presenting spatial data, processing and analysing spatial data, data input and editing, can be developed (TG: O'Connor, 2007). More recently, articles have reflected on changes in GIS functionality, such as ArcGIS Online Story Maps (TG: Walshe, 2016) and WebGIS (TG: West and Horswell, 2018). Both these latter articles, written by a teacher educator (TG: Walshe, 2016) and academic geographers (TG: West and Horswell, 2018) link these developments with implications for teacher and student learning in geography and example classroom activities.

Within professional literature in the last decade there has also been a shift towards focusing on GIS in relation to a particular geographical focus, such as the sustainability of the Olympic stadium (TG: Martin, 2011) and glacial landforms (TG: Clark et al., 2018). This would appear to mirror the turn in academic literature to foreground how GIS can be used to develop students' geographical knowledge and thinking (Fargher, 2017; Walshe, 2018; Healy and Walshe, 2020). This professional discourse also provides insights that address Rød et al.'s (2010) question: 'How can GIS help meet curricular goals?' (p.34). Walshe (2018) has similarly highlighted that teachers must grasp how GIS can contribute to a high-quality geography curriculum and become a tool to develop students' capacity to think geographically. Articles written by geography teachers about their practice can help render this invisible form of curriculum thinking visible. Other articles have more explicitly placed a focus on how teachers perceive that GIS provides a greater capacity for their students' geographical thinking. For example, Heath (TG: 2016) highlighted that by using GIS students were able to examine the vulnerability of the local population around and the global consequences of a volcanic eruption, while Walshe (TG: 2016) explicates how Story Maps can enable students to understand different spatial scales when exploring the incidence of disease at the local, national and international scale. Further, Cadwallader (PG: 2006) discussed how GIS enabled her Year 4 pupils to develop 'a sense of place that had both a spatial and temporal dimension' (p.25). These vignettes could provide a mechanism for other teachers to reflect on how GIS could enhance their students' geography learning, which might be a motivating factor for teachers to sustain their own personal learning and implementation of GIS (Höhnle et al., 2016).

Trafford's (TG: 2017) focus on developing and refining a department's approach to embedding GIS within their geography teaching across four years, and Heath's (TG: 2015; 2016) two articles, illustrate that developing classroom practice and sharing through professional discourse is a long-term endeavour. This is also considered by O'Connor (TG: 2007; 2009) who exemplifies how his teaching practice has evolved with the advancing GIS technologies. O'Connor (2009) shifts from using a desktop GIS to WebGIS and provides a clear exposition of the different opportunities arising because of the development of GIS technologies. This all appears to illustrate Collins and Mitchell's (2019) recommendations that a long-term approach is needed to professional learning around GIS for teachers, as in these instances embedding GIS takes place over several years. For example, Trafford (TG: 2017) outlines that the first two academic years were about experimenting, the next year was about trialling and refining, and the fourth year provided an opportunity for innovating and embedding GIS. Whilst the authors of these articles do not explicitly reference the TPACK model (Mishra and Koehler, 2006), it appears that TPACK in relation to GIS is being treated as a conglomerate (Angeli and Valanides, 2009). These geography teachers do not immediately use GIS technology in their teaching, but spend time assimilating new GIS technologies, identifying where there is opportunity for GIS to be integrated into the geography curriculum and then transforming their teaching practices (Oda et al., 2019). Oda et al. (2019) emphasise that this epistemological view of TPACK means that it takes longer for teachers to assimilate new technology and for GIS to be implemented within the classroom.

This form of professional discourse is not just about a contribution to professional knowledge, but also serves to provide a stimulus for other geography teachers to reflect on their practice. For example, Heath (TG: 2016) poses several questions for departmental discussion, including: 'To what extent can using GIS activities in lessons enhance learning beyond textbook-based maps? How can GIS activities be incorporated across different topics? Are there opportunities for further developing students' spatial literacy skills?' (p.114). Such questions have the potential to illustrate the value of GIS, which has been emphasised as an important motivator for teachers' professional development around GIS by Höhnle et al. (2016) and Walshe (2018).

It appears that engagement with such professional literature for teachers can provide a form of 'continuous follow-up' (Collins and Mitchell, 2019, p.132), because over time teachers are able to read and reflect upon their own use of GIS as they encounter each new article. It can also be used to induct trainee teachers into the professional knowledge base around GIS, as the curricular and pedagogical decision-making of teachers is illuminated. This moves beyond teachers developing awareness of GIS and towards an understanding of the role GIS can play within their and others' geography teaching.

What is revealed about teachers' professional practice in relation to GIS?

Within *Teaching Geography*, 13 articles referenced GIS in relation to fieldwork; these were mostly concerned with using GIS for data collection and data presentation. This would seem to be indicative of quite a limited view of how GIS can be marshalled as part of fieldwork investigations and enquiries. However, two articles did show there was recognition of the capacity for GIS to be used as a way of preparing for fieldwork or accessing secondary data to support students' geographical enquiries. This has been perceived to be a marker for successful engagement with GIS, as this use of GIS 'guides the questions students are able to ask', allowing students 'to formulate and answer questions' (Healy and Walshe, 2020, p.15).

GIS was also mentioned frequently in relation to geographical skills that students should acquire as part of their geographical education (e.g. PG: Palfrey, 2006; TG: May, 2014; PG: Owens, 2018). In some places, this reference was just in passing to GIS skills as specified in the Geography National Curriculum (DfE, 2013) and in current revisions of GCSE and A-Level specifications. The notions of GIS skills in the context of these curriculum documents were often not subject to discussion or problematisation, or considered for their potential to develop students' geographical knowledge and thinking (Walshe, 2018). For example, in the 2017 AS examiners' reports, use of GIS was highlighted as a geographical skill in need of attention (TG: Palôt et al., 2018). It would seem apt to reflect on whether Wood's (TG: 2006) concerns that 'geographical experiences are not well examined through the use of formal examination, including skills relating to fieldwork and GIS' (p.137) were well grounded. Roberts (TG: 2017) foregrounds that GIS can contribute to a powerful geographical education where students are helped 'to make sense of the world' (p.8). Moving beyond seeing GIS as a means to an end (Walshe, 2018), Rackley (TG: 2018) illuminates where GIS can support the development of students' geographical questioning skills at Key Stage 3. GIS has also been recognised as a key element in a geography teacher's repertoire as part of delivering a 'varied lesson diet' (TG: Selmes, 2012, p.72). While HMI and National Advisor for Geography, Iwaskow (TG: 2013) highlighted that whilst GIS was perceived to be part of an effective geography education, 'most schools are … failing to meet statuary requirements' (p.55) in relation to GIS. Within a small-scale study, Healy and Walshe (TG: 2019) found that only two of 15 students knew what GIS was when they started their A-Level geography course, which, alongside reflections from Palôt et al. (TG: 2018), means there is some evidence to tentatively suggest that we have not yet overcome this under-utilisation of GIS in school geography.

Simmons (TG: 2016) explicated the value of GIS for visualising data by foregrounding students' comments on this: 'The use of ArcGIS Online helped me to see the dramatic retreat of real-life glaciers' (p.17). Watt (TG: 2006) explicitly evaluates her first use of GIS with a Year 8 class in relation to the outcomes for student learning. This form of reflection is notable in relation Höhnle et al. (2016), as it highlights where teachers are reflecting on how GIS is successfully supporting their students' learning.

GIS was also signposted in relation to career opportunities for geographers (TG: Amis, 2012). For example, Frontani (TG: 2004) highlighted how GIS is being used in geography at university level through the discussion of undergraduate student projects at Elon university, and Hennig (TG: 2018) heralded the role GIS had in making 'the cartographer's life considerably easier' (p.66). These examples outline where stakeholders in geography education are contributing to professional discourse that illuminates the role of GIS in geography more broadly. Pollard and Hesslewood (TG: 2015) outline how ubiquitous GIS is within both public and private sectors, but more explicitly extend this by demonstrating how they have developed meaningful opportunities to engage students with how GIS is being used in the real world, such as introducing how hazard mapping can inform government policy. GIS was also referred to as a process of enquiry by Taylor (TG: 2008), who suggests that 'in some enquiry sequences, you might want to have a particular focus on processes of enquiry – how geographers find out about the world, perhaps using maps, GIS or fieldwork' (p.53). Healy (TG: 2018) articulated that GIS provided a way for students to better understand their A-Level place study and gave scope for students to engage with academic geographers' research, which enabled them to appreciate the role of GIS in geographical research undertaken in the academy. These examples illustrate that within professional discourse GIS is not just perceived as a skill (Walshe, 2018), but is being drawn upon within curriculum development and teaching for the insights it can provide into academic and real-world geography for students.

Finally, across both journals GIS was referenced as something that departments should hold in mind when evaluating their practice (TG: Hopkin, 2006) and setting a vision (TG: Ward, 2017). For example, Baston (TG: 2016) illustrates this in relation to the development of a flipped learning model, where their next steps were to 'integrate more sophisticated GIS activities and fieldwork' (p.71). Further, Willy (PG: 2017) suggests that primary teachers could 'challenge [themselves] to do things that do not necessarily come easily to [them] – map work and GIS for example' (p.7). Glesinger et al. (TG: 2017) also highlighted that 'supporting colleagues' CPD had led to the creation of GIS resources which [they] have shared through [their] own networks and through the Geographical Association' (p.53). This signals that there is sharing of both practice and resources in relation to GIS within the geography subject community through various avenues.

Conclusion

This chapter explores how professional discourse within the geography education community opens up a way of understanding how teachers are marshalling GIS within their professional practice. It is necessary to highlight that this work is limited in its scope, as it solely focuses on the professional journals of *Primary Geography* and *Teaching Geography*. There are other professional activities and discourses that could be examined, such as master's level practitioner research, summaries of lectures and workshops presented at the Geographical Association Annual Conference, and published teaching resources and textbooks.

Professional discourse appears to render visible the repertoire of experience held within the geography subject community. Persaud (2011) has argued that *Teaching Geography* has played a significant role in making and remaking a legitimate version of school geography and encouraging teachers to achieve professional autonomy. Within a subject community, there should be greater consideration for professional agency, where the professional judgement of the subject-specialist teachers is acknowledged. Whilst it is proposed that individuals' past experience enables them to achieve professional agency (Priestley et al., 2015), I would argue that individual teachers, especially beginning teachers, need to benefit from the past experience found within their subject community: this would mean beginning teachers are less likely to be constrained in their curriculum making by their first-hand experience.

Whilst there are very few explicit references to academic research within geography teachers' professional discourse in *Teaching Geography*, it appears that there have been similar shifts in what has been explored by geography teachers as in the academic community. As GIS technology rapidly develops, it seems ever more important for researchers in the academy to pay attention to how the concerns of practitioners evolve. Professional discourse is also particularly significant in relation to GIS, because it both illuminates teachers' motivations for using GIS and has the capacity to influence other teachers' dispositions towards GIS. As Brooks (2018) highlights, within a practice-orientated field there is a need to foreground the dialogic relationship between theory and practice. Professional discourse in relation to GIS illuminates how practitioners have found ways to overcome some of the challenges acknowledged in academic literature, such as rendering visible curricular purposing of GIS and enabling other practitioners to understand the long-term commitment to professional learning needed to embed GIS within their geography teaching. I would argue, like Brooks (2018), that this dialogue has to be two-way, and ought to recognise the value of insights from practitioners.

Within a subject community of teachers, Ellis (2007) highlights that newcomers have the 'potential to stimulate the generation of new knowledge through a creative disruption of existing practices' (p.452), as 'practice itself is in motion' (Lave and Wenger, 1991, p.34). Walshe (2017) has already stressed that trainee teachers can themselves be empowered as 'experts' and go on to develop department practice with GIS. This exemplifies that, as new technologies develop and create different opportunities and challenges, there is a greater necessity to capitalise upon the collective role of practitioners and researchers in developing geography teachers' professional practice in relation to GIS.

References

Amis, K., 2012. Finding the big question. *Teaching Geography*, 37 (1), pp.10–11.
Angeli, C., and Valanides, N., 2009. Epistemological and methodological issues for the conceptualization, development, and assessment of ICT–TPCK: Advances in technological pedagogical content knowledge (TPCK). *Computers and Education*, 52, pp.154–168.
Baker, T.R., 2015. WebGIS in education. In O.M. Solari, A. Demirci and J. Schee (eds), *Geospatial Technologies and Geography Education in a Changing World*. London: Springer, pp.105–115.
Baker, T.R., Battersby, S., Bednarz, S.W., Bodzin, A.M., Kolvoord, B., Moore, S., Sinton, D. and Uttal, D., 2015. A research agenda for geospatial technologies and learning. *Journal of Geography*, 114 (3), pp.118–130.
Baston, J., 2016. A flipped learning model. *Teaching Geography*, 41 (2), pp.70–71.
Bednarz, S.W. and Ludwig, G., 1997. Ten things higher education needs to know about GIS in primary and secondary education. *Transactions in GIS*, 2, pp.123–133.
Bednarz, S.W. and van der Schee, J., 2006. Europe and the United States: The implementation of geographic information systems in secondary education in two contexts. *Technology, Pedagogy and Education*, 15 (2), pp.191–205.
Brooks, C., 2018. Insights on the field of geography education from a review of master's level practitioner research. *International Research in Geographical and Environmental Education*, 27 (1), pp.5–23.
Brooks, C., Brant, J., Abrahams, I. and Yandell, J., 2012. Valuing initial teacher education at Master's level. *Teacher Development: An International Journal of Teachers' Professional Development*, 16 (2), pp.285–302.

Cadwallader, T., 2006. A new look at GIS in the primary classroom. *Primary Geography*, 59, pp.24–25.

Clark, C.D., Ely, J.C., and Doole, J., 2018. Glacial landforms: A teaching resource in maps and GIS. *Teaching Geography*, 43 (2), pp.76–79.

Collins, L. and Mitchell, J.T., 2019. Teacher training in GIS: What is needed for long-term success? *International Research in Geographical and Environmental Education*, 28 (2), pp.118–135.

Counsell, C., 2012. 'The other person in the room': A hermeneutic-phenomenological inquiry into mentors' experience of using academic and professional literature with trainee history teachers. In M. Evans (ed.), *Teacher Education and Pedagogy: Theory, Policy and Practice*. Cambridge: Cambridge University Press, pp.134–181.

Counsell, C., Evans, M., McIntyre, D. and Raffan, J., 2000. The usefulness of educational research for trainee teachers' learning. *Oxford Review of Education*, 26 (3), pp.467–482.

Cox, S. and Graham, C.R., 2009. Using an elaborated model of the TPACK framework to analyze and depict teacher knowledge. *TechTrends*, 53 (5), pp.60–69.

Denscombe, M., 2007. *The Good Research Guide: For Small-Scale Social Research Projects* (3rd edn). Milton Keynes: Open University Press.

Department for Education (DfE), 2013. *Geography Programmes of Study: Key Stage 3*. Available at: www.gov.uk/government/uploads/system/uploads/attachment_data/file/239087/SECONDARY_national_curriculum_-_Geography.pdf [Accessed 11 May 2020].

Doering, A., Veletsianos, G., Scharber, C. and Miller, C., 2009. Using the technological, pedagogical, and content knowledge framework to design online learning environments and professional development. *Journal of Educational Computing Research*, 41 (3), pp.319–346.

Ellis, V., 2007. Taking subject knowledge seriously: From professional knowledge recipes to complex conceptualizations of teacher development. *The Curriculum Journal*, 18 (4), pp.447–462.

Fargher, M., 2017. GIS and the power of geographical thinking. In C. Brooks, G. Butt, and M. Fargher (eds), *The Power of Geographical Thinking*. London: Springer, pp.151–164.

Fargher, M., 2018. WebGIS for geography education: Towards a GeoCapabilities approach. *ISPRS International Journal of Geo-Information*, 7 (3), pp.1–15.

Freeman, D., Green, D. and Hassell, D., 1994. A guide to geographic information systems (GIS). *Teaching Geography*, 19 (1), pp.36–37.

Frontani, H., 2004. Experiential learning: Bring the geography of Africa to life. *Teaching Geography*, 29 (3), pp.132–134.

Geographical Association (GA), 2011. *GIS: Training the Trainer*. Available at: www.geography.org.uk/gtip/gis/#top [Accessed 11 May 2020].

Glesinger, K., Sloneczny, L. and Wall, S., 2017. Leading geography. *Teaching Geography*, 42 (2), pp.52–53.

Gómez Trigueros, I.M., 2018. New learning of geography with technology: The TPACK model. *European Journal of Geography*, 9 (1), pp.38–48.

Healy, G., 2018. Using local organisations and geographical scholarship to support A-Level place studies. *Teaching Geography*, 43 (1), pp.13–15.

Healy, G. and Walshe, N., 2019. Leveraging real-world geographers and GIS: Relevance, inspiration and developing geographical knowledge. *Teaching Geography*. 44 (2), pp.52–55

Healy, G. and Walshe, N., 2020. Real-world geographers and geography students using GIS: Relevance, everyday applications and the development of geographical knowledge. *International Research in Geographical and Environmental Education*, 29 (2), pp.178–196.

Heath, R., 2015. A world-wide geographical investigation using online GIS. *Teaching Geography*, 40 (3), pp.118–120.

Heath, R., 2016. Developing an explosive scheme of work. *Teaching Geography*, 41 (3), pp.112–114.

Hennig, B., 2018. Worldmapper: Rediscovering the world. *Teaching Geography*, 43 (2), pp.66–68.

Höhnle, S., Fögele, J., Mehren, R. and Schubert, J., 2016. GIS teacher training: Empirically based indicators of effectiveness. *Journal of Geography*, 115 (1), pp.12–23.

Hong, J.E. and Stonier, F., 2015. GIS in-service teacher training based on TPACK. *Journal of Geography*, 114 (3), pp.108–117.

Hopkin, J., 2006. Writing a geography SEF. *Teaching Geography*, 31 (1), pp.33–36.

Iwaskow, I., 2013. Geography: A fragile environment. *Teaching Geography*, 38 (2), pp.53–55.

Jo, I., 2016. Future teachers' dispositions toward teaching with geospatial technologies. *Contemporary Issues in Technology and Teacher Education*, 16 (3), pp.310–327.

Kelly, A., 2006. HOTShot GIS. *Primary Geography*, 67, pp.28–29.

Kidman, G. and Palmer, G., 2006. GIS: The technology is there but the teaching is yet to catch up. *International Research in Geographical and Environmental Education*, 15 (3), pp.289–296.

Koehler, M. and Mishra, P., 2009. What is technological pedagogical content knowledge (TPACK)? *Contemporary Issues in Technology and Teacher Education*, 9 (1), pp.60–70.

Lave, J. and Wenger, E., 1991. *Situated Learning: Legitimate Peripheral Participation*. Cambridge: Cambridge University Press.

Martin, F., 2011. Beyond 2012: Using GIS to investigate the sustainability of the Olympic stadium. *Teaching Geography*, 36 (1), pp.22–23.

May, C., 2014. Planning a new Key Stage 3. *Teaching Geography*, 39 (3), pp.100–101.

Mishra, P. and Koehler, M.J., 2006. Technological pedagogical content knowledge: A framework for teacher knowledge. *Teachers College Record*, 108 (6), pp.1017–1054.

O'Connor, P., 2007. Progressive GIS. *Teaching Geography*, 23 (3), pp.147–150.

O'Connor, P., 2009. The application of UK web-based GIS in geography teaching. *Teaching Geography*, 34 (1), pp.19–25.

Oda, K., Herman, T. and Hasan, A., 2019. Properties and impacts of TPACK-based GIS professional development for in-service teachers, *International Research in Geographical and Environmental Education*. doi: doi:10.1080/10382046.2019.1657675/.

Owens, P., 2018. A chameleon subject with a rigorous heart. *Primary Geography*, 97, pp.30–32.

Palfrey, D., 2006. Taking a different review. *Primary Geography*, 61, pp.32–34.

Palôt, I., Hore, H., Oakes, S. and Digby, B., 2018. How can the AS examiners' reports help improve your students' performance at A-Level? *Teaching Geography*, 41 (1), pp.9–10.

Persaud, I., 2011. The daring discourses of *Teaching Geography? Geography*, 96 (3), pp.137–142.

Pollard, G. and Hesslewood, A., 2015. A more 'authentic' geographical education. *Teaching Geography*, 40 (1), pp.11–13.

Priestley, M., Biesta, G. and Robinson, S., 2015. *Teacher Agency: An Ecological Approach*. London: Bloomsbury.

Rackley, S., 2018. How to ... develop (independent investigation) questioning skills at home. *Teaching Geography*, 43 (1), pp.11–12.

Ricker, B. and Thatcher, J., 2017. Evolving technology, shifting expectations: Cultivating pedagogy for a rapidly changing GIS landscape. *Journal of Geography in Higher Education*, 41 (3), pp.368–382.

Roberts, M., 2017. Geographical education is powerful if.... *Teaching Geography*, 42 (1), pp.6–9.

Rød, J.K., Larsen, W. and Nilsen, E., 2010. Learning geography with GIS: Integrating GIS into upper secondary school geography curricula. *Norsk Geografisk Tidsskrift – Norwegian Journal of Geography*, 64 (11), pp.21–35.

Selmes, I., 2012. The risk in excellent teaching. *Teaching Geography*, 37 (2), pp.71–72.

Shulman, L.S., 1986. Those who understand: Knowledge growth in teaching. *Educational Researcher*, 15 (2), pp.4–14.

Simmons, G., 2016. Teaching glaciation. *Teaching Geography*, 41 (1), pp.16–17.

Slater, F. (ed.), 1995. *Reporting Research in Geography Education* (Monograph No. 2). London: Institute of Education, University of London.

Taylor, L., 2008. Key concepts and medium term planning. *Teaching Geography*, 33 (2), pp.50–54.

Trafford, R., 2017. How to ... integrate GIS effectively into the curriculum. *Teaching Geography*, 42 (3), pp.106–107.

van Manen, M., 1990. *Researching Lived Experience: Human Science for an Action Sensitive Pedagogy*. London, ON: Althouse Press.

Walshe, N., 2016. Working with ArcGIS Online Story Maps. *Teaching Geography*, 41 (3), pp.115–117.

Walshe, N., 2017. Developing trainee teacher practice with geographical information systems (GIS). *Journal of Geography in Higher Education*, 41 (4), pp.608–628.

Walshe, N., 2018. Spotlight on ... Geographical information systems for school geography. *Geography*, 103 (1), pp.46–49.

Ward, L., 2017. Starting out as a leader in geography. *Teaching Geography*, 42 (3), pp.103–104.

Watt, S., 2006. Talk the talk: Mapping mobile phone masts with GIS. *Teaching Geography*, 31 (1), pp.30–32.

West, H. and Horswell, M., 2018. GIS has changed! Exploring the potential of the ArcGIS Online. *Teaching Geography*, 43 (2), pp.22–24.

Willy, T., 2017. Sustaining primary geography. *Primary Geography*, 93, pp.6–7.

Wood, P., 2006. Internal assessments in the pilot GCSE: Styles and implications. *Teaching Geography*, 31 (3), pp.136–137.

Chapter 10

Empowering geography teachers and students with geographical knowledge
Epistemic access through GIS

Mary Fargher and Grace Healy

Introduction

Despite the development of WebGIS, there are questions over whether geospatial technologies remain underutilised within school geography (Höhnle et al., 2016; Walshe, 2017; Collins and Mitchell, 2019); however, little is known about the extent to which GIS is embedded within geographical education. This chapter investigates one possible route forward in strengthening the connections that a digitally conscious teacher could make between the use of GIS and the development of geographical knowledge. To do so, it evaluates the use of four different WebGIS applications in developing students' geographical knowledge: Esri ArcGIS Online to study earthquake patterns; Esri Survey 123 to carry out an environmental quality survey in London's King's Cross; QGIS Cloud to undertake a geographical enquiry on fracking in Dorset; and Esri StoryMaps to study the 'hottest place on Earth' – the Danakil Depression in Ethiopia. Through these examples, the argument presented is that in order for these new and exciting WebGIS tools to contribute to students' geographical education, geography teachers need to consider more explicitly their role in developing students' capacity to 'think geographically' (Fargher, 2018).

GIS in geography education

Internationally, the benefits of GIS for supporting spatial thinking (National Research Council and Geographical Sciences Committee, 2008), mapping and representing geographical phenomena (Lei et al., 2009) and supporting enquiry-based learning (Favier and Van der Schee, 2012) are well-documented. Enquiry is acknowledged as a key feature of high-quality geography education in several countries (Roberts, 2013; Ferretti, 2018) and a means by which GIS can be successfully employed in geography education (Walshe, 2018). GIS can provide access to geographical data that enables students to develop their own questions and places them in a position to interrogate 'the whys of where' through high-quality mapping and analysis of data (Kerski, 2003; 2018; Healy and Walshe, 2020; Pike, 2021). When used in these ways, GIS can be powerful medium for teaching and learning about geographical topics and can bring to life issues of real-world relevance to students' lives (Fargher and Rayner, 2012; Pike, 2021).

In England, recent curriculum and examination reforms have led to a more prominent role for GIS in geography education. Students at Key Stage 3 (aged 11–14) are expected to 'use Geographical Information Systems (GIS) to view, analyse and interpret places and data' (DfE, 2013, p.3). Building on this, students at Key Stage 4 (aged 14–16) are now required to be able to use GIS 'to obtain, illustrate, analyse and evaluate geographical information'

(DfE, 2014a, p.5). Finally, at Key Stage 5 (aged 16–18) students are required to understand the nature of geospatial technologies that are used to collect, analyse and present geographical data and 'demonstrate an ability to collect and to use digital, geo-located data' (DfE, 2014b, p.13). As outlined by Walshe (2018), such requirements mean that teachers can be drawn into a mechanistic, box ticking exercise in which students 'do GIS' alongside other skills (such as statistics), rather than using GIS to support them to think geographically' (p.47). Mitchell (2020) provides evidence for this, as he shares the words of one geography teacher to illustrate how teachers are concerned over covering aspects of the curriculum: 'I'm worried about how we're going to do (GIS) if Ofsted come in and ask how we do this. The plan is to put it in the "investigating my world"' (Mitchell, 2020, p.130). Despite this, as highlighted by Healy (2021), there are geography teachers that are focusing on how GIS can be used to develop students' geographical thinking; for example, Heath (2016) ensured that students' developed an understanding of the importance of the geographical concept of scale in relation to the consequences of a volcanic eruption.

WebGIS

Despite the educational potential of GIS, some teachers still associate the use of GIS with a steep initial learning curve. In the past, the use of industry-orientated GIS and poor access to geospatial data often dissuaded teachers from attempting to use GIS. The shift to WebGIS has removed many logistical obstacles (Baker, 2015). In comparison with earlier forms of GIS, WebGIS is much more intuitive with user-friendly interfaces. Web-based mapping platforms, such as ArcGIS Online and QGIS Cloud, are much more commonplace in schools. In 2018, the Royal Geographical Society shared the news that 'over 1700 UK schools have taken up their [ArcGIS Online's] free subscription, with over 57,000 pupils actively using the software as part of their geographical studies' (RGS, 2018). In the next sections, we consider more explicitly how particular features of WebGIS can be beneficial for school geography.

Intuitive mapping

One of the most significant innovations associated with WebGIS has been the introduction of intuitive mapping. Intuitive mapping involves GIS that has inbuilt capacity to direct a user to the most appropriate method of map-making and analysis. For example, smart mapping is a capacity built into ArcGIS Online which allows novice users to produce professional quality maps quickly and easily regardless of their knowledge of GIS (Fargher, 2018).

Mobile geospatial technologies

A major innovation in the use of WebGIS in schools has been the development of mobile applications such as Survey123 for ArcGIS. This includes cloud-based editable feature services (EFS) which enables teachers to design fieldwork with mobile applications 'where they can control how, where and what their students collect whilst out in the field' (Fargher, 2018, p.4).

Open source data

The impact of open source data on the use of geospatial applications in schools has become increasingly significant, as geography teachers can now access a vast range of multi-scale geospatial data. WebGIS tools also encourage more user-contributed data, for example through citizen science, crowdsourcing and volunteered geographical information (VGI). EDINA's (2019) DigiMap OpenStream, a specialised application programming interface, provides free web map services (WMS) to the academic community and schools, (Fargher, 2018). Within ArcGIS Online, the Living Atlas also provides teachers and students with immediate access to a collection of data sets that can be explored and interpreted to support students' studies in geography (West and Horswell, 2018).

Customised web mapping applications

Also significant are customised web mapping applications, such as Esri StoryMaps,[1] that combine interactive web map templates with a variety of digital online content, such as text, photographs or videos. Teachers or students can use a range of map templates to build story maps from an online gallery or can create their own maps within the ArcGIS Online platform. The advantages for teaching and learning geography with customised web mapping applications wide-ranging, including their use in telling the story of places, events, issues or patterns in a geographical context (Walshe, 2016).

Powerful disciplinary knowledge

In this next section, we begin by articulating the context of the 'knowledge-turn' and why this is significant for realising the potential of GIS within geographical education. The concept of powerful disciplinary knowledge (PDK) has origins in the work of Michael Young (2008) who argues that all young people regardless of background are entitled to 'powerful knowledge'. Young (2009) has argued that it is distinct from common sense knowledge, because it is systematic (related to each part of a discipline) and specialised (developed by specialists within defined fields of expertise). This has been developed within the geography education community in the context of GeoCapabilities and PDK has been described by Lambert (2016) as including: 'the acquisition of deep descriptive "world knowledge"', 'the development of relationship thinking that underpins geographical thought' and 'a propensity to apply the analysis of alternative social, economic and environment futures to particular place contexts' (pp.404–405). Additionally, Maude (2016) has identified powerful knowledge offered through geography by considering the 'intellectual power it gives to those who have it' (p.10). A capability approach expresses 'what people are able to "do" or "be"' (Bustin, 2019, p.100), and, therefore, we argue that Maude's (2016) typology (Table 10.1) can be used to complement this approach. Yet, there appears to have been limited consideration of GIS within the context of Geo-Capabilities (e.g. Lambert et al., 2015; Bustin, 2019) beyond the development of PDK vignettes within a story map[2] and Fargher's insights into how PDK could be developed through a 'curriculum artefact' (Fargher, 2018).

Maude's (2016) typology has been applied to analysing the geographical knowledge that could be developed through use of a curriculum artefact within ArcGIS Online (Fargher, 2018) and through an empirical account of the geographical knowledge developed by A-Level geographers as part of fieldwork enquiry using GIS (Healy and Walshe, 2020). As the central purpose of this chapter is to critically consider the extent to which GIS used in schools can contribute to developing students' powerful geographical knowledge, in the next sections we use Maude's (2016) typology of powerful geographical knowledge to

Table 10.1 Maude's (2016) typology of powerful geography knowledge adapted by Lambert and Solem (2017, p.11).

Type	Characteristic
1 Knowledge that provides students with 'new ways of thinking about the world'	Using 'big ideas' such as: • place • space • environment • interconnection These are meta-concepts that are distinguished from substantive concepts, like *city or climate*

2 Knowledge that provides students with powerful ways of analysing, explaining and understanding	Using ideas to:
	• *analyse*
	  - e.g. place; spatial distribution
	• *explain*
	  - e.g. hierarchy; agglomeration
	• *generalise*
	  - e.g. models, such as push-pull models of migration; and laws

3 Knowledge that gives students some power over their own knowledge	To do this, students need to know something about the ways knowledge is developed and tested in geography. This is about having an answer to the question: 'How do you know?' This is an underdeveloped area of geographical education, but is a crucial aspect of 'epistemic quality' (Hudson, 2016)
4 Knowledge that enables young people to follow and participate in debates on significant local, national and global issues	School geography has a good record in teaching this knowledge, partly because it combines the natural and social sciences, and the humanities. It also examines significant issues such as: food, water and energy security; climate change; development
5 Knowledge of the world	This takes students beyond their own experience – the world's diversity of environments, cultures, societies and economies. In a sense, this knowledge is closest to how geography is perceived in the popular imagination. It contributes strongly to a student's 'general knowledge'

examine the extent to which different approaches to GIS can be used within school geography to assure epistemic access. A full description of how each approach has the potential to develop powerful geographical knowledge is given within a corresponding table. We will draw out key points in the main text, exploring the first two examples in detail and then using the final two examples to exemplify pertinent points that relate to the uniqueness of that approach to GIS.

Analysing earthquake patterns in Esri ArcGIS Online

This first example uses data from the United States Geological Survey (USGS) Earthquakes Hazards Program[3] to collect, map and analyse earthquake patterns. Data is stored on the website and can then be transferred into a GIS platform, such as ArcGIS Online or QGIS Cloud. The smart mapping capability available in ArcGIS Online means that CSV (comma separated values) data in spreadsheet format can be easily dragged, dropped and displayed on a base map. Once the data is dropped onto the map it is displayed as shown in the example which illustrates the relationships between earthquake magnitude and depth (Figure 10.1).

Table 10.2 summarises the kinds of powerful geographical knowledge about earthquake patterns that can be developed in ArcGIS Online. In this scenario, it provides capacity to develop students' knowledge of geography's key concepts, such as place, space and interconnection (Brooks, 2018: type 1 powerful knowledge). Understanding of these concepts underpins access to other types of geographical knowledge as they shape the geographical lens teachers and students can use in geography lessons; however, this conceptual knowledge is also continually reshaped as students come to appreciate geographical concepts as 'sites of contestation ... [that] ... possess multiple meanings' (Lambert and Morgan, 2010, p.xi). Drawing on the USGS data can enable students to analyse the distribution of earthquakes, and can be used alongside other geographical evidence to explain why earthquakes have a greater impact in some countries than others (type 2 knowledge). It is also feasible that changing earthquake patterns mapped over time could be used to understand historical tectonic activity and test seismic gap theory to predict where earthquakes may happen in the future (type 2 and type 3 knowledge). The power of this has been illuminated by Hawley's (2018) consideration of the mapping of earthquake epicentres, depths and magnitudes in Mexico between 1900 and 2010. Hawley (2018) highlights that the historical pattern of earthquakes does not fit with the classic 'model' subduction zone, where you would expect to see a 'linear belt of significant earthquakes running parallel to the (offshore) trench' (p.33). This would help students understand the complexity of theory and reality in the real world, and emphasises the importance of the local geographical context in mediating the uniqueness of tectonic activity within a place and, therefore, taking account of the seismotectonic setting (type 1 knowledge). In Mexico, it is through combining this mapping with seismic tomography that the complexity of tectonic activity can be understood (Hawley, 2018). This also gives students a better understanding of how GIS can be used by geographers to

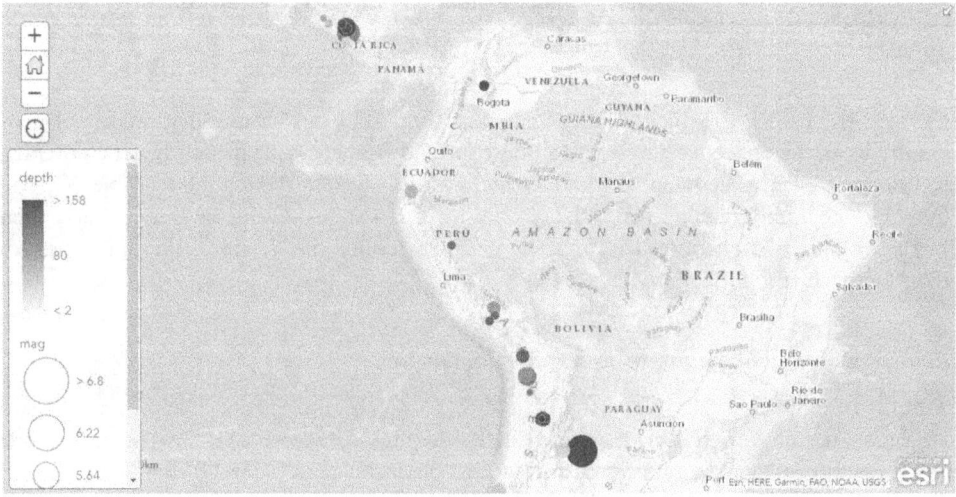

Figure 10.1 Mapping relationships between earthquake magnitude and depth in ArcGIS Online.

Table 10.2 Powerful geographical knowledge developed through analysing earthquake patterns with ArcGIS Online (Example 1).

Type according to Maude (2016)	Powerful geographical knowledge developed through GIS
1 Knowledge that provides students with 'new ways of thinking about the world'	Knowledge of key geographical concepts such as place, space, interconnection and risk
2 Knowledge that provides students with powerful ways of analysing, explaining and understanding	Powerful ways to analyse, explain and understand earthquake patterns that involves: • Mapping and visualising major earthquakes, tectonic plate boundaries and seismic gaps along faults • Mapping and identifying the most powerful earthquakes that occur along convergent plate boundaries • Visualising and making connections between the characteristics of the most destructive earthquakes (e.g. earthquakes occurring at shallow depth, and near populated areas)
3 *Knowledge that gives students some power over their own knowledge*	Students ask their own spatial questions, visualise earthquake spatial data and perform spatial analysis to map, analyse and explain earthquake patterns. Students can understand how GIS can be used by geographers to develop hazard mapping
4 Knowledge that enables young people to follow and participate in debates on significant local, national and global issues	Students develop their knowledge of how earthquake impact and recovery vary across different countries. For example: • Mapping and analysing the characteristics of different earthquakes to understand the importance of different factors (e.g. the development of the country, investment in earthquake-proofing buildings, vulnerability to multiple hazards)
5 Knowledge of the world	Students study locations beyond their direct experiences

Note: The type is italicised if not all characteristics suggested in Table 10.1 are achievable directly through the use of GIS.

develop hazards mapping (Pollard and Hesslewood, 2015: type 3 knowledge). Exploration of USGS earthquake data can then allow students to ask their own spatial questions and explore any significant relationships between the data sets available (type 3 knowledge). Further to this, for particular events, students could map shaking intensity as a way to understand how ground conditions mediate the intensity of an earthquake and, therefore, the distribution of impacts from a seismic event (Hawley, 2018: type 2 knowledge). Finally, Fearnley (2021) highlights the role of Twitter in providing a source of data for geographers; for example, geo-tagged insights could be used within ArcGIS Online to develop students' understanding of the spatial and temporal extent of earthquakes (e.g. Earle et al., 2011).

Collecting and analysing environmental quality data with Esri Survey 123

Teachers can create their own customised surveys within Esri's Survey 123, or can enable students to develop their own surveys for use in geography fieldwork. In this second example, Figure 10.2 shows the base map used in conjunction with Survey 123 in a field study

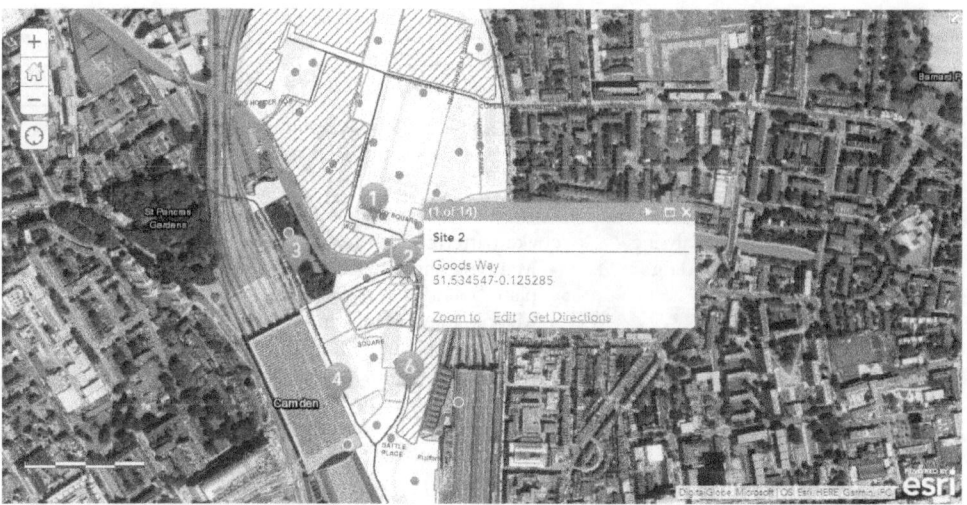

Figure 10.2 Identifying sites in King's Cross for environmental quality fieldwork data with Survey 123.

carried out by UCL Institute of Education Secondary Geography PGCE[4] students investigating environmental quality in London King's Cross.

As Table 10.3 shows, there are a number of key elements that can be identified to support students in developing powerful geographical knowledge by using Survey 123 to collect, map and analyse data on environmental quality. Students are able to ask their own spatial questions, collect and visualise their own data, and perform spatial analysis to help them analyse and explain variations in environmental quality, which provides them both with power over their own fieldwork and the opportunity to understand how different methods can be employed with geographical research (type 3 knowledge). Through Survey 123, students can format questions in different ways and develop surveys that are underpinned by different constructions of the concept of place, which in turn enables them to develop a more nuanced understanding of the concept (type 1 knowledge). For example, a phenomenological approach to place might mean that students would explore peoples' perceptions of environmental quality and what influences these (Cresswell, 2015; Rawling, 2018). Students could also draw upon air pollution monitoring records (including hourly air pollution indexes) from King's College London's Environment Research Group,[5] which could be compared with other methods of collecting data on air pollution, or be correlated against people's perceptions of air quality; this has the potential to advance students' understanding of geographers' and scientists' research in this area (type 3 knowledge). There is also considerable scope to develop geographical knowledge that enables students to partake in debates about significant and contentious local issues, such as urban redevelopment (type 4 knowledge); as highlighted by Pike (2021), this could lead to students being able to contribute their informed perspectives to stakeholders and develop their own understanding so they can be geographically informed 'active citizens'.

This use of WebGIS not only has a role in developing school students' geographical knowledge, but was also used in this scenario to develop teachers' 'experience of learning in the field' and provide an opportunity for them to 'begin to organise learning opportunities

Table 10.3 Powerful geographical knowledge developed through collecting and analysing environmental quality data with Esri Survey 123 (Example 2).

Type according to Maude (2016)	Powerful geographical knowledge developed through GIS
1 Knowledge that provides students with 'new ways of thinking about the world'	Knowledge of key geographical concepts, such as place, space and environment
2 Knowledge that provides students with powerful ways of analysing, explaining and understanding	Powerful ways to analyse, explain and understand variations in environmental quality in London King's Cross, that involves: • Collecting and storing environmental data regarding criteria such as access to open space, noise, traffic, litter, aesthetic quality, air pollution • Mapping and analysing data collected with Survey 123 in ArcGIS Online • Comparing contemporary and historical ArcGIS maps to identify changes over time • Producing and analysing graphs and other infographics in ArcGIS Online
3 *Knowledge that gives students some power over their own knowledge*	Students ask their own spatial questions, visualise environmental quality data and perform spatial analysis to analyse and explain variations in environmental quality. Students can draw on secondary data sets and develop an understanding of the different methods of geographical research in this area
4 Knowledge that enables young people to follow and participate in debates on significant local, national and global issues	Students develop their knowledge of contrasts between land use within London King's Cross and can consider different stakeholders' points of view when debating significant inner-city redevelopment issues
5 Knowledge of the world	This type of geospatial analysis takes students beyond their own experience and into the complexity of the real world to study London King's Cross

Note: The type is italicised if not all characteristics suggested in Table 10.1 are achievable directly through the use of GIS.

in the field for others' (Lambert and Reiss, 2016, p.33). Furthermore, Brooks (2010) has highlighting the importance of this for beginning teachers:

> The ability can be simulating for both the teacher and the students as they generate new understandings about the Geography and the environment in situ. The ability to create such learning experiences outside the classroom is, arguably, one of the most challenging dimensions of a geography teacher's work. (p.123)

In the context of this field study, beginning teachers are learning to navigate how to create meaningful learning in the field and how GIS can be used to effectively support this. This enables trainee geography teachers to develop self-efficacy with GIS within the context-specific scenarios that they are likely to encounter in their teaching practice, so they are likely to be able to use geospatial technologies with students as part of fieldwork in a way that enhances geographical learning. Whilst a collective field study for trainee teachers might be helpful in developing their own confidence in using GIS, Walshe (2017) and Collins and Mitchell (2019) highlight the importance of embedded use of GIS, so this becomes part of a geography teacher's repertoire of practice. Walshe (2017) emphasises within ITE it is necessary to engage with school mentors and support the development of their practice with GIS, so that mentors and trainees can collaboratively work together on using GIS with students.

Investigating fracking in QGIS Cloud

This third example examines QGIS materials for teachers with EDINA's (2019) Mapstream for Schools WMS to investigate the issue of fracking. The resource includes a range of data sources that can be brought together in GIS including Ordnance Survey map data and Open StreetMap and Google Maps satellite and Street View imagery to study the issue of fracking for Wareham in Dorset.

Table 10.4 illustrates that, by using a range of map, satellite and Street View layers in QGIS Cloud students can visualise and analyse the impact of fracking in the area (type 2 knowledge). Controversial issues in geography are, as Mitchell (2018) argues, part of the 'challenge of the "super-complex" and "radically unstable" world that students grapple with' (p. 228); this uncertainty can hinder learners, but Barnett (2011) argues that this can be overcome if students have a 'willingness to be changed as a result of one's learning' (p.11). Within the digital world, it is ever more important that students develop their geographical knowledge through critically engaging with geographical sources; this should thereby build

Table 10.4 Powerful geographical knowledge developed through investigating fracking in QGIS Cloud (Example 3).

Type according to Maude (2016)	*Powerful geographical knowledge developed through GIS*
1 Knowledge that provides students with 'new ways of thinking about the world'	Knowledge of key geographical concepts, such as place, space, environment and sustainability
2 Knowledge that provides students with powerful ways of analysing, explaining and understanding	Powerful ways to analyse, explain and understand fracking at Wytch Farm, Wareham in Dorset, that involves: • Using Open StreetMap, Google Maps and Street View data to map the fracking sites and areas of outstanding natural beauty (AONB) • Mapping land use in the area • Mapping conflicts of interest • Measuring distances from fracking sites to residential areas and AONB
3 *Knowledge that gives students some power over their own knowledge*	Students ask their own spatial questions, visualise spatial data and perform spatial analysis to analyse and explain different land uses and to gauge environmental impacts and economic benefits
4 Knowledge that enables young people to follow and participate in debates on significant local, national and global issues	Students use GIS to carry out a geographical enquiry on fracking, investigating the issue from a range of stakeholder perspectives using geo-tagged sources
5 Knowledge of the world	This type of geospatial analysis takes students beyond their own experience to study a complex issue such as fracking

Note: The type is italicised if not all characteristics suggested in Table 10.1 are achievable directly through the use of GIS.

their capacity to grapple with the potential misrepresentation around significant geographical issues, such as energy security (Roberts, 2021; type 3 and type 4 knowledge). Students can also begin to draw on local evidence from a range of sources, including maps, photographs, and geo-tagged sources, to consider fracking from a range of stakeholder perspectives; this enables them to move towards an understanding of how place identity or attachment might underpin those perspectives (type 4 knowledge). As highlighted by Roberts (2013), it is necessary in the planning stage to think carefully about how understanding of geography's key concepts can be used and developed; for example, an explicit focus upon people's attachment to place could develop students' conceptual understanding of place.

Using an Esri StoryMap to develop knowledge of Danakil, Ethiopia

This fourth example uses an Esri StoryMap. In this example Garry Simmons, Head of Geography at Wilmington Grammar School for Girls, UK, produced the StoryMap 'Journey into Danakil: Hottest Place on Earth' to show the route of an expedition from Ethiopia's Mekele Airport into the heart of the Danakil Depression, drawing on a range of multimedia content to enhance this geographical story. The StoryMap has been developed to enable students to ask and answer a range of geographical questions about the extreme environments encountered along the way and to provide information on a host of significant physical and human geography features (Simmons, 2015).

The use of Esri StoryMaps to develop powerful geographical knowledge is significant in a number of ways (Table 10.5). In this example, the teacher has used the resource to introduce students to geographies they could not experience directly and that can stimulate their geographical imaginations of far-off places such as the Danakil Depression in Ethiopia (type 5 knowledge), and can foreground significant issues facing this area, so students can develop

Table 10.5 Powerful geographical knowledge developed using an Esri StoryMap to study Danakil, Ethiopia (Example 4).

Type according to Maude (2016)	Powerful geographical knowledge developed through GIS
1 Knowledge that provides students with 'new ways of thinking about the world'	Knowledge of key geographical concepts, such as place and environment
2 Knowledge that provides students with powerful ways of analysing, explaining and understanding	Powerful ways to analyse, explain and understand earthquake patterns, including: • Mapping and visualising the Danakil Depression, Ethiopia • Identifying and analysing key physical features • Identifying and analysing key human characteristics
3 *Knowledge that gives students some power over their own knowledge*	Students ask their own spatial questions, visualise spatial data and perform a number of spatial queries about the main geographical features of Danakil
4 Knowledge that enables young people to follow and participate in debates on significant local, national and global issues	Students develop their knowledge of geographical issues arising from drought, challenging terrain, and, in some areas, poorly developed infrastructure and political unrest.
5 Knowledge of the world	This takes students beyond their own experience to study distant places

Note: The type is italicised if not all characteristics suggested in Table 10.1 are achievable directly through the use of GIS.

an informed perspective of geographical challenges arising from drought, and in some areas, lack of infrastructure and political unrest (type 4 knowledge). Using GIS in this way can support students' place-based learning as Walshe (2018) has suggested and would provide a model that students could then use to develop their own StoryMaps (type 3 knowledge). Walshe (2017) found that trainee teachers were more confident with using StoryMaps than other aspects of GIS, and so there might be greater capacity to draw on this use of GIS for developing students' geographical knowledge. Though, within ArcGIS Online, there are new applications such as Community Analyst that enable access to up-to-date geographical data, which has the potential to empower students to be able to develop rich and nuanced insights about any chosen place (Hall and Sawle, 2018), and so it is important that teachers do not just remain focused on the aspects of GIS they are most comfortable with.

Conclusion

This chapter began by referring to the well-established advantages of using GIS in geography education with its record of supporting the development of spatial skills and as a medium for geographical enquiry. It was suggested that one of the key issues for geography education with regards to the use of GIS is the under-theorised relationship between its use and the development of powerful geographical knowledge. Yet, as we have shown in this chapter, there appear to be a wealth of advantages for geographical education in:

- Using GIS to map and analyse spatial patterns of earthquakes that cannot be recognised outside of a GIS (Example 1).
- Using mobile applications such as Esri Survey 123 to collect and analyse geospatial data about environmental quality in the field (Example 2).
- Using cloud-based mapping platforms such as QGIS Cloud to construct multi-layered geographical enquiries about controversial issues such as fracking (Example 3).
- Using customised web map applications such as Esri StoryMaps to support in-depth place-based learning (Example 4).

Maude's (2016) typology has provided a useful tool to consider where GIS enables epistemic access, but its limitations have implications for future research and practice. Whilst this chapter has gone some way to explicating how using GIS can provide access to geographical knowledge, there needs to be a more developed theorisation of the role this knowledge plays within geography curricula design for teachers and the extent to which the use of GIS to enable epistemic access can be embedded throughout students' school geography journey.

We found, like Bouwmans and Béneker (2018), that Maude's (2016) typology underemphasises the importance of the connections that exists between the types of geographical knowledge. This was particularly of important in relation to Type 1 and Type 3 knowledge. As highlighted within the first example, a geographical concept can underpin the development of other aspects of geographical knowledge, but is also developed through other geographical knowledge. Within the examples chosen here, the geographical concepts of place and space have appeared to dominate; however, it might be necessary to more deeply consider the opportunities for development in relation to other concepts (for example, globalisation and risk; Clifford et al., 2008). The four examples of WebGIS all provide some scope to address type 3 knowledge; however, this would often be reliant upon the subject expertise of the teacher and the capacity to reach beyond what is immediately available within GIS. This is significant given that this is an area that is under-theorised more broadly within school geography, as outlined by Firth (2015), Lambert and Solem (2017: see Table 10.1) and Healy (forthcoming). Further to this, Lambert and Solem (2017) argue that type 3 knowledge is 'crucial, for an essential element of powerful knowledge is an understanding of its dynamic nature' (p.10); therefore, an absence or lack of type 3 knowledge becomes a potential barrier to students' entitlement to powerful geographical knowledge.

This chapter has shown that in-depth consideration of the role of GIS on the development of powerful geographical knowledge can empower teachers' curriculum thinking (Fargher, 2017; Walshe, 2018). The removal of the barriers to GIS in schools – cost, complexity, access – has provided a real watershed for those teachers now incorporating GIS use into their classrooms. New developments in intuitive mapping, easier-to-use mobile applications, open source data and bespoke customised mapping applications are providing greater opportunities for geospatial information to be drawn upon within school geography. However, geography teachers need to be digitally conscious to 'be aware of the different discourses which are being represented within [GIS]' (Fargher, 2013, p.238). Given that geographers have emphasised the extent to which 'powerful digital technologies can reproduce the lies and politics' that have existing within cartography for decades (Swords et al., 2019, p.142), this emphasises the importance of engagement with metadata when using GIS to develop students' geographical knowledge (Fargher, 2017). To further capitalise upon these innovations, there needs to be a focus on how GIS can be embedded within geography teachers' repertoire of practice, so that it becomes part of their curriculum making. Without this, we are not just denying students their curriculum entitlement to GIS, but also the power GIS offers in terms of epistemic access in school geography.

Notes

1 https://storymaps.arcgis.com/en/
2 https://esriukeducation.maps.arcgis.com/apps/MapJournal/index.html?appid= 828f082b0ed444b8b92d4502bb20d3b9
3 https://earthquake.usgs.gov
4 The PGCE is a one-year higher education course in England, Wales and Northern Ireland which provides training in order to allow graduates to become teachers within maintained schools. As part of it, trainee teachers usually spend time in both university and school settings.
5 www.londonair.org.uk/LondonAir/Default.aspx

References

Baker, T.R., 2015. WebGIS in education. In O. Muñiz Solari, A. Demirci and J.A. van der Schee (eds), *Geospatial Technologies and Geography Education in a Changing World*. Dordrecht: Springer, pp.105–115.

Barnett, R., 2011. Learning about learning: A conundrum and a possible resolution. *London Review of Education*, 9 (1), pp.5–13.

Bouwmans, M. and Béneker, T., 2018. Identifying powerful geographical knowledge in integrated curricula in Dutch schools. *London Review of Education*, 16 (3), pp.445–459.

Brooks, C., 2010. Learning to teach geography. In R. Helibronn and J. Yandell (eds), *Critical Practice in Teacher Education: A Study of Professional Learning*. London: Institute of Education, University of London, pp.114–124.

Brooks, C., 2018. Insights on the field of geography education from a review of master's level practitioner research. *International Research in Geographical and Environmental Education*, 27 (1), pp.5–23.

Bustin, R., 2019. *Geography Education's Potential and the Capabilities Approach: GeoCapabilities and Schools*. London: Palgrave Macmillan.

Clifford, N., Holloway, R., Rice, S., Valentine, G., Holloway, S.L. and Rice, S.P. (eds), 2008. *Key Concepts in Geography*. London: Sage.

Collins, L. and Mitchell, J.T., 2019. Teacher training in GIS: What is needed for long-term success? *International Research in Geographical and Environmental Education*, 28 (2), pp.1–18.

Cresswell, T., 2015. *Place: A Short Introduction*. Oxford: Blackwell.

Department for Education (DfE), 2013. *Geography Programmes of Study: Key Stage 3*. Available at: https://assets.publishing.service.gov.uk/government/uploads/system/uploads/attachment_data/file/239087/SECONDARY_national_curriculum_-_Geography.pdf [Accessed 31 May 2019].

Department for Education (DfE), 2014a. *GCSE Subject Content*. Available at: https://assets.publishing.service.gov.uk/government/uploads/system/uploads/attachment_data/file/301253/GCSE_geography.pdf [Accessed 31 May 2019].

Department for Education (DfE), 2014b. *GCE A Level Subject Content*. Available at: https://assets.publishing.service.gov.uk/government/uploads/system/uploads/attachment_data/file/388857/GCE_AS_and_A_level_subject_content_for_geography.pdf [Accessed 31 May 2019].

Earle, P.S., Bowden, D.C. and Guy, M., 2011. Twitter earthquake detection: Earthquake monitoring in a social world. *Annals of Geophysics*, 54 (6), pp.708–715.

EDINA, 2019. *Digimap OpenStream*. Available at: https://openstream.edina.ac.uk/ [Accessed 31 May 2019].

Fargher, M., 2013. A Study of the Role of GIS in Constructing Relational Place Knowledge through School Geography Education. PhD thesis, Institute of Education, University of London.

Fargher, M., 2017. GIS and the power of geographical thinking. In C. Brooks, G. Butt and M. Fargher (eds), *The Power of Geographical Thinking*. Cham: Springer International, pp.151–164.

Fargher, M., 2018. WebGIS for geography education: Towards a GeoCapabilities approach. *ISPRS International Journal of Geo-Information*, 7 (3), pp.1–15.

Fargher, M. and Rayner, D., 2012. United Kingdom: Realizing the potential for GIS in the school geography curriculum. In A.J. Milson, A. Demirci and J.J. Kerski (eds), *International Perspectives on Teaching and Learning with GIS in Secondary Schools*. Dordrecht: Springer. pp.299–304.

Favier, T.T. and Van der Schee, J.A., 2012. Exploring the characteristics of an optimal design for inquiry-based geography education with geographic information systems. *Computers and Education*, 58 (1), pp.666–677.

Fearnley, F. 2021. Social media as a tool for geographers and geography educators. In N. Walshe and G. Healy (eds), *Geography Education in the Digital World*. London: Routledge. pp.75–85.

Ferretti, J., 2018. The enquiry approach in geography. In M. Jones and D. Lambert (eds), *Debates in Geography Education*. London: Taylor and Francis, pp.115–126.

Firth, R., 2015. Constructing geographical knowledge. In G. Butt (ed.), *MasterClass in Geography Education: Transforming Teaching and Learning*. London: Bloomsbury, pp.53–66.

Hall, K. and Sawle, J., 2018. *ArcGIS for Schools Update*. Esri UK. Available at: www.slideshare.net/Esri_UK/update-on-arcgis-online-for-schools-smart-education-schools-ac18 [Accessed 3 January 2020].

Hawley, D., 2018. On shaky ground: The physical facts of recent earthquake events in Mexico. *Teaching Geography*, 43 (1), pp.32–35.

Healy, G., 2021. Insights from professional discourse on GIS: A case for recognising geography teachers' repertoire of experience. In N. Walshe and G. Healy (eds), *Geography Education in the Digital World*. London: Routledge, Chapter 9.

Healy, G., forthcoming. A call to view disciplinary knowledge through the lens of geography teachers' professional practice. In M. Fargher, D. Mitchell and E. Till (eds), *Recontextualising Geography for Education*. Cham: Springer.

Healy, G. and Walshe, N., 2020. Real-world geographers and geography students using GIS: Relevance, everyday applications and the development of geographical knowledge. *International Research in Geographical and Environmental Education*, 29 (2), pp.178–196.

Heath, R., 2016. Developing an explosive scheme of work. *Teaching Geography*, 41 (3), pp.112–114.

Höhnle, S., Fögele, J., Mehren, R. and Schubert, J., 2016. GIS teacher training: Empirically based indicators of effectiveness. *Journal of Geography*, 115 (1), pp.12–23.

Hudson, B., 2016. Didactics. In D. Wyse, L. Hayward, and J. Pandya (eds), *The Sage Handbook of Curriculum, Pedagogy and Assessment*. London: Sage, pp.107–124.

Kerski, J.J., 2003. The implementation and effectiveness of geographic information systems technology and methods in secondary education. *Journal of Geography*, 102 (3), pp.128–137.

Kerski, J.J., 2018. Explaining the whys of where at TED. *GeoNet: The Esri Community*, 23 March. Available at: https://community.esri.com/community/education/blog/2018/03/23/explaining-the-whys-of-where-at-ted [Accessed 3 January 2020].

Lambert, D., 2016. Geography. In D. Wyse, L. Hayward and J. Pandya (eds), *The Sage Handbook of Curriculum, Pedagogy and Assessment*. London: Sage, pp.391–407.

Lambert, D. and Morgan, A., 2010. *Teaching Geography 11–18: A Conceptual Approach*. Maidenhead: Open University Press.

Lambert, D. and Reiss, M.J., 2016. The place of fieldwork in geography qualifications. *Geography*, 101 (1), pp.28–34.

Lambert, D. and Solem, M., 2017. Rediscovering the teaching of geography with the focus on quality. *Geographical Education*, 30, pp.8–15.

Lambert, D., Solem, M. and Tani, S., 2015. Achieving human potential through geography education: A capabilities approach to curriculum making in schools. *Annals of the Association of American Geographers*, 105 (4), pp.723–735.

Lei, P.L., Kao, G.Y.M., Lin, S.S. and Sun, C.T., 2009. Impacts of geographical knowledge, spatial ability and environmental cognition on image searches supported by GIS software. *Computers in Human Behavior*, 25 (6), pp.1270–1279.

Maude, A., 2016. What might powerful geographical knowledge look like? *Geography*, 101 (2), pp.70–76.

Mitchell, D. 2018. Handling controversial issues in geography. In M. Jones and D. Lambert (eds), *Debates in Geography Education*. London: Taylor and Francis, pp.224–236.

Mitchell, D., 2020. *Hyper-Socialised: How Teachers Enact the Geography Curriculum in Late Capitalism*. London: Routledge.

National Research Council and Geographical Sciences Committee, 2008. *Learning to Think Spatially*. Washington, DC: National Academies Press.

Pike, S. 2021. GIS for young people's participatory geography. In N. Walshe and G. Healy (eds), *Geography Education in the Digital World*. London: Routledge, pp.117–128.

Pollard, G. and Hesslewood, A., 2015. A more 'authentic' geographical education. *Teaching Geography*, 40 (1), pp.11–13.

Rawling, E., 2018. Place in geography: Change and challenge. In M. Jones and D. Lambert (eds), *Debates in Geography Education*. London: Routledge, pp.49–61.

RGS, 2018. *Society Supports Esri in Promoting GIS to Schools*. Available at: www.rgs.org/geography/news/society-supports-esri-in-promoting-gis-to-schools/ [Accessed 3 January 2020].

Roberts, M., 2013. *Geography through Enquiry*. Sheffield: Geographical Association.

Roberts, M., 2021. Geographical sources in the digital world: Disinformation, representation and reliability. In N. Walshe and G. Healy (eds), *Geography Education in the Digital World*. London: Routledge, pp.53–64.

Simmons, G., 2015. *Journey into the Danakil: Hottest Place on Earth: A Story Map*. Available at: www.arcgis.com/apps/MapJournal/index.html?appid=739bbe980baf4878916a99eb918b1eae [Accessed 3 January 2019].

Swords, J., Jeffries, M., East, H. and Messer, S., 2019. Mapping the city: Participatory mapping with young people. *Geography*, 104 (3), pp.141–147.

Walshe, N., 2016. Using ArcGIS Online Story Maps. *Teaching Geography*, 41 (3), pp.115–117.

Walshe, N., 2017. Developing trainee teacher practice with geographical information systems (GIS). *Journal of Geography in Higher Education*, 41 (4), pp.608–628.

Walshe, N., 2018. Spotlight on ... Geographical information systems for school geography. *Geography*, 103 (1), pp.46–49.

West, H. and Horswell, M., 2018. GIS has changed! Exploring the potential of the ArcGIS Online. *Teaching Geography*, 43 (2), pp.22–24.

Young, M., 2008. From constructivism to realism in the sociology of the curriculum. *Review of Research in Education*, 32 (1), pp.1–28.

Young, M. 2009. What are schools for? In H. Daniels, H. Lauder and J. Porter (eds), *Knowledge, Values and Educational Policy: A Critical Perspective*. London: Routledge, pp.10–18.

Chapter 11

GIS for young people's participatory geography

Susan Pike

Introduction

Thirty years ago Goodchild (1988) predicted that digital technologies would have the 'potential to stimulate a revolution in cartography' (p.318). This prediction has come true as geospatial technologies, combined with a range of social media platforms, have transformed how we negotiate the world as citizens. Today millions of events, large and small, are captured in place and scale via technology (Shin and Bednarz, 2019, p.viii), with the young people[1] we teach being an integral part of this process. At the same time, within geography lessons, digital technologies are used for both teaching and learning, on whiteboards, tablets and out in the field (Fargher, 2017a).

This chapter explores how young people's potential and existing expertise in using such technology can be harnessed for valuable participatory geographical learning with a view to educating 'a whole child for the purposes of allowing the child to thrive in a world with unprecedented changes to environment, society and technology' (Chang and Kidman, 2019, p.1). Whilst such experiences can also occur in other subjects, notably citizenship, languages and history, the focus here is on the potential of geography. For the purpose of this chapter, participatory geographical learning activities are characterised as enquiry approaches within which young peoples' interests and curiosity are a starting point for learning, but where teachers shape how the learning process proceeds (Roberts, 2013; Pike, 2016). Participatory geographical learning activities broadly have four main underpinning components:

- Enquiry approaches to learning, where young peoples' interests and curiosities are the starting point for learning, but teachers shape how this learning can proceed (Roberts, 2013; Pike, 2016).
- Geographical content that is relevant to young people but that also takes them beyond their everyday experiences to increase their capabilities (Catling, 2003; Young, 2008).
- Lessons that involve doing geography through interactions with places and people, supported by teachers (Heffron and Downs, 2012).
- Possibilities of actions by young people, such as informed decisions, arising from the enquiry learning process in lessons (Pike, 2016; Shin and Bednarz, 2019), whilst recognising that action may be limited by political structures (Wynne-Jones et al., 2015).

Within participatory geographical learning, the inextricable link between geography and citizenship is always evident. Shin and Bednarz (2019) consider this relationship, suggesting that geography identifies and analyses relationships among concepts central to geographical enquiry, such as space, place and environment, whereas citizenship applies geographical information and concepts in order to make informed decisions for the common good generally within justice frameworks. With this in mind, at the heart of this chapter is the consideration of how, as educators, we can draw together young people's social, geographical and technological experiences to enhance geographical learning in schools.

Young people's everyday use of geospatial technologies

For many young people, use of technology, particularly on their phones, is a common experience. An example of this is through the phenomenally successful application Snapchat, with its embedded feature 'Snapmaps' which allows users to see where their Snapchat contacts are, share their current location, and view Snaps from nearby Snapchat users or users at a specific event or location. Although not originally conceived as such, according to Snapchat's own blog (2018), location is a key part of how users interact with the social media platform. This is illustrated through a range of 'spatial' functionality; for example, Snaps (cartographically viewing snaps from a particular place) or Stories (viewing stories posted in particular places). There appears to be some disagreement about the value of such technology and concerns associated with its use, as reflected in the media as well as in research work. Whilst parents have been encouraged to be alert to problems such as cyberbullying, sexting and exposure to inappropriate content (O'Keeffe and Clarke-Pearson, 2011), Orben et al. (2019) have concluded that social media use on mobile phones was not a strong predictor of life satisfaction.

As well as any potential social benefits, spatial-based applications such as Snapchat provide a way to teach about geospatial technologies, including geographical information systems (GIS), in schools. Turner (2006) describes 'neogeographers' as non-experts who contribute to geographical techniques and tools for personal and community activities despite no training. Through their phones, neogeographers offer volunteered geographic information (VGI) (Goodchild, 2007) by way of spatially-referenced applications; for example, a geotagged photograph might be actively uploaded onto Snapchat or Facebook. This provides the opportunity to teach the difference between VGI and contributed geographical information (CGI) (Goodchild, 2007; Blatt, 2015), and the ethical use of georeferenced data by both the initial application (such as Snapchat) and the subsequent purchasers of that information. CGI only applies when there is an opportunity to opt out of the data production process, such as background use of data on Find My iPhone (Blatt, 2015). Geography teachers might explore use of VGI within geographical decision-making. One such example is the work of the Humanitarian Open Street Map Team on humanitarian action and community development through open mapping (HOSM, 2019); this includes mapping of 8,000km of roads and 300,000 buildings in the Democratic Republic of Congo to help with the tracking and management of the outbreak of Ebola. Through teaching students not just about how data is collated, analysed and possibly shared by companies using GIS, they can appreciate the processes behind the compilation of GIS maps and understand more about what happens to the data they share. Social media can itself be used as a tool for geographers (see Fearnley, 2021) and, therefore, this also can serve as basis for understanding geographical research.

Young people and local places

Local places are key to our daily experiences, including homes, schools, workplaces, clubs and neighbourhoods (Lynch, 1960). Across decades of research it is evident that the range of experiences young people have in their locality are particularly important to them, including their current wellbeing, self-esteem and identity, as well as cognitive and affective development (Hart, 1979; Pike, 2011; Schlemper et al., 2018). Furthermore, young people also have direct and indirect experience of other places through family connections and travels. Through geospatial technologies, these links are made more tangible using maps, photographs and video footage; all of these can be developed and explored in school contexts.

Research exploring students' perceptions of and experiences in their locality has found they use and value a range of places (Tunstall et al., 2004; Pike, 2011): those that are available, those that they like, and those where they can be with friends or perhaps alone (Chawla, 2002; Schlemper, et al., 2018). As Lynch (1960), Hart (1979), and Massey (2005) note, there are different ways to view places, and research with children has consistently found their views of places differ from that of adults (e.g. Martin, 1999; Pike, 2011). Anyone who has walked along a street with a child will appreciate the imagination with which they are able to see places and features; for example, the scary 'face' in a tree, or the way they notice the smallest insects on a wall. Places vary temporally in their significance to children; a wind-swept bridge to the sea becomes more important in the summer as it is such a social place, to meet and jump into the water! Today, with a mobile phone, such experiences of places merge into digital worlds, as locations of friends can be tracked in order to coordinate the meetup at that bridge, and experiences can be mapped, photographed, commented on and shared through applications such as Snapchat and Facebook. Through the use of such applications, young people now often live in both real and virtual worlds; the implications of this technological change for children's geographies is explicated by Hammond (2021). From Pike's work with young people (Pike, 2011; 2016; 2020a; 2020b), it is evident that they hold experiences that help them understand geographical information systems, such as:

- The personal maps of the locality with 'overlays', such as places they remember when they were younger, friends' homes, as well as other significant places such as school, sports clubs, etc.
- The GIS adults use, such as seeing and hearing sat navs in parents' cars to watching teachers use Google Maps at school.
- The location-based technology they use, for example, playing map-based computer games such as Minecraft, and becoming users of applications such as Snapchat and Pokémon Go.

Like any digital GIS, young people's experiences of geographical information systems are not stable, they change temporally; Giddens (1991) noted the intricacies of this, describing how 'time-space paths criss-cross in the contexts of daily life, constitute that life as "normal" and predictable' (p.126). Such experiences provide a rich basis for geography and citizenship lessons in schools; as students know about places, they appreciate the differing uses and views of them, and may have gained first-hand appreciation of local geographical processes, such as changes in land use and the creation of new features. They also have the ability to identify local issues and suggest solutions to address them (Ramezani and Said, 2013), although these may contrast with adults' ideas (Laughlin and Johnson, 2011). As they move on, they can also consider why and how their locality is as it is, and consider possible futures (Hicks, 2007; 2014). As McMahon et al. (2018) note, 'participation and active citizenship are not just about involvement in decision-making but about social and civic participation – being a "part" of, not "apart" from, civic society' (p.127). It has been argued that young people are made 'invisible' as a result of limited community participation in urban planning (Alarasi et al., 2016). Therefore, the combination of using the locality for both geography and citizenship lessons has the potential to provide opportunities for community participation for young people (Pike, 2020b).

Connecting to the local with GIS

Young people, from preschool through to secondary age, are capable of using technology and taking part in participatory, enquiry-based geography lessons with GIS. In Ireland, students have been using GIS in schools for nearly 20 years, although this has generally been for competitions such as the extremely popular national Young Scientist Competition. There are also numerous examples of research that has illustrated the power of such combinations of learning with technology. For example, Danby et al. (2016) found that the use of GIS technology through Google Earth broadened preschool teachers' understandings of children's socio-spatial practices. They report that children's virtual explorations of their local school and community sites on Google Earth enhanced teachers' appreciation of children's capabilities, both with regards to use of technology and their learning about and understanding of the local community (Danby et al., 2016). This suggests that even with young children, the use of the locality, enhanced by geospatial technologies, can be beneficial to children's learning. In the UN Convention on the Rights of the Child (UNCRC), Article 12 recognises children's right to express views freely and have them taken seriously (UN, 1989). Working within such a rights-based participatory perspective, Schlemper et al. (2019) explored use of a sequence of enquiry-based geography lessons within which students themselves identified local issues to consider, including crime, housing, and youth employment opportunities, with GIS to support them. They found that a range of learning took place, including development of the students' awareness of their community, improvements in their spatial thinking, and progress in their understanding of GIS. Further, Egiebor and Foster (2019) found that use of Esri's Story Maps allowed students to identify geographic and cultural connections between places depicted on Story Maps and their lives, thereby connecting them to geography beyond the classroom. Walshe (2016) also suggests that Story Maps allow students to explore place in more detail through interrogating a wide range of geographical information, including the integration of local data sets (e.g. crime data).

Where students are inspired to ask geographical questions about local issues, GIS can help them analyse spatial data quickly, so that they can reach the stage of suggesting solutions to issues more effectively. In Northern Ireland, ArcGIS was used to bring schools together to map children's perceptions of their local places, with the aim of breaking down barriers between students living segregated lives due to their religion (Esri, 2019; Kenny, 2019). Learning with GIS can go beyond traditional school curricula, as young people can share their findings with local decision makers, such as local councillors, engineers and planners (Pike, 2020b). For example, Berglund and Nordin (2007) found during a local participation project that planners were more responsive to young peoples' ideas about urban planning because of the credibility given to them through using GIS.

In Ireland, national policy initiatives and the school curriculum all include participatory geography in both schools or the wider community. This is particularly true of the National Strategy on Children and Young People's Participation in Decision-Making, which 'focuses on the everyday lives of children and young people and the places and spaces in which they are entitled to have a voice in decisions that affect their lives', and specifically states that 'children and young people will have a voice in decisions made in their local communities' (DCYA, 2015, p.3). The National Digital Strategy (DCENR, 2013) makes specific reference to digital mapping including community mapping, although falls short of suggesting this should be a school-based activity. The Primary School Curriculum introduction (4/5 years to 12/13 years[2]), has a number of key principles that embrace participatory geographical learning, including 'the child is an active agent in his or her learning' and 'the child's immediate environment provides the context for learning' (DES/NCCA, 1999b, p.8). The curriculum also recognises the role of ICT as a resource which considerably enriches the teaching and learning of different aspects of the curriculum (DES/NCCA, 1999a; 1999b). Whilst geography in the primary curriculum is under review, digital spatial technologies are specifically mentioned in the new mathematics curriculum (DES/NCCA, 1999b; 2018). There are opportunities to base much of the teaching at Junior Certificate level on participatory geography as the syllabus is underpinned with geo-literacy and sustainability; the opportunities for participatory geography are numerous. There are many opportunities to use GIS within the Junior Cycle (12/13 to 16/17 years old) to support the development of geographical skills; while there are frequent references to technology, geospatial technologies are not explicitly mentioned (DES/NCCA, 2017). The classroom-based assessments offer similar opportunities as they are on Geography in the News and 'My Geography', an enquiry into a geographical aspect of the local area. In the current Leaving Certificate specification (18/19 years old), GIS is explicitly included within the application of geographical skills as a tool that should be used to study geographical problems: 'GIS, as a specialised investigative tool, can be used to combine data sources in the study of particular areas or geographical problems' (DES/NCCA, 2003, p.20). The curriculum documentation goes on to suggest that this could include local issues such as 'the destruction of archaeological sites' (DES/NCCA, 2003, p.20). The new Leaving Certificate, due in the coming years, is more likely to specifically refer to GIS, as more schools now have broadband access and, since 2018, all schools have free access to ArcGIS (Kelly, 2018).

Drawing on young peoples' geographies to enhance geographical learning

The following examples of students' participatory mapping with GIS are from a primary and secondary school in Ballygall in Dublin. *Baile na nGall* (Irish) was settled by Vikings in the 11th century, and later by the Cambro-Normans, ultimately becoming a townland divided between the rural villages of Finglas and Glasnevin. Initial discussions with the students from both schools indicated they knew a lot about their locality, as identified in previous research (e.g. Pike, 2011), and their understanding of their locality was complex and evolving. As a starting point, they reflected characteristics of Lynch's (1960) five elements of mental maps:

- 'Paths' or routes in a locality, such as the stretch (a local term for the River Tolka) and bus routes to homes of others.
- 'Edges' as real or imagined boundaries between one place and another, such as the Royal Canal, which forms a boundary between districts.
- 'Districts' or specific areas of a locality with official or local names, such as Ballygall (a suburb of the city) and the North Circular (a local term for the area each side of the North Circular Road).

- 'Nodes' or points within a locality, for example shops (such as Centra) or risky places (such as a busy road junction).
- 'Landmarks', such as Johnstown Park (a local park).

Projects were then undertaken with students at the two schools (Sacred Heart Boys' National School and St Kevin's College) to develop students' understanding of and engagement with their local area using GIS.

Sacred Heart Boys' National School

At the start of their final primary school year, the Sacred Heart pupils, aged 11/12 years old, were asked to locate and describe signage and other features in the Irish language in their local area. Following this, the boys were asked to suggest questions about their locality; these included:

- What was the first location in Finglas?
- Why is Finglas so old?
- Why is Finglas/Ballygall where it is?
- How much did it cost to build Finglas?
- How did some things change and why?
- How did the houses look in olden times?
- How long did the stream stretch?

As is often the case with student constructed enquiry questions, their questions also revealed insights into their geographies and identities, such as their fascination with their locality and community (Pike, 2016). The children were supported to investigate some of these questions by a group of Bachelor of Education student teachers from the local university who were specialising in geography education. Groups of five children were paired with two student teachers to support pupil-led walks around the local area where places of particular significance to the children were visited; they were also directed to find something 'old', 'new', 'unusual' and 'interesting'. Photography was used by the boys as a way of actively engaging with the landscapes around them (Hall, 2015). From the student teachers' perspective this project was very positive: 'The boys provided lots of knowledge regarding the changes in the community over the last few years. They were engaged and actively involved, they had a positive attitude and were eager to learn more'. The pupils then worked with the student teachers again, this time with paper maps from 1900 and the present day in order to answer their own questions about how the area had changed. The boys also created maps using GIS via Google Maps. The photographs the students had taken were all added to an online photograph hosting site (Flickr), so they could access and choose their own or others' photographs to map. The pupils did this on one class map, accessed through a class Gmail account which meant they could easily see the places chosen by other groups. Interestingly they could also see different views of the same place, enabling the boys to gain another opportunity to engage with their locality and explore what their peers had noticed.

To extend this further, historical photographs were used to help students understand how their locality had changed. This was similar to Davies et al. (2019), who reported on a task where undergraduate students revisited locations in Berlin where photographs were taken in the 1980s to recreate these photographs and share them via Instagram, with findings posted up on Twitter. Such a 'rephotography' task, where 'the photograph offers a visual record of "moments" that can be measured against future change' (Sanders, 2007, p.185), would enable primary-aged students to engage in the process of 'actively looking' (Hall, 2009) at their present-day locality and understand how their area had changed over time.

St Kevin's College

The students at St Kevin's College were in their fourth year of secondary school (aged 15–17 years old); this is referred to as the Transition Year between two sets of state examinations in Ireland (Junior Certificate and Leaving Certificate). In most schools this will involve students taking core school subjects but carrying out more project-based work in other subjects and work experience. At St Kevin's the students, who were all boys, had the opportunity to carry out an open-ended enquiry on an issue of their choice in geography. Like the boys at Sacred Heart, the first activity was to generate ideas for their investigations by taking part in a walk around their locality. Over a number of lessons, the students then learnt the basic functionality of ArcGIS Online, such as how to create their own maps, and began to think about questions they could research using it. The projects were diverse but were mainly issues-based, such as:

- Gang land crime: In recent years in Dublin there had been a number of shootings arising from rival gangs. The groups wondered whether the levels of crime had increased or if they were being reported more often.
- Accident black spots: As pedestrians and cyclists, one group were aware of certain places in their locality where there had been a number of accidents. They were curious as to why some places were accident black spots and other places were not.
- Locations of schools and fast food shops/shops with hot food counters: In Ireland most convenience stores have a 'fast food' counter selling chips, sausages, etc. The group felt this was intentional on the part of the store owners. The boys had noted there was such a shop near their own school and wondered if this was the case for other schools.

To draw out the value of these issues-based enquiries for students' geographical education, we will focus upon a group of students that explored how the location of facilities enabling environmentally responsible action differed across localities in their city. The students were already conscious of the geographical concept of 'spatial justice', contending with whether there was 'fair and equitable distribution in space of socially valued resources and opportunities to used them' (Soja, 2009, p.2), as they noted differences between parts of the city in terms of wealth. They also noted that there appeared to be more recycling banks, bike stands and car charging points in some parts of the city than others, and sensed these facilities were located in wealthier areas. Using GIS, the students were able to examine the spatial patterns of the provision of these facilities. The students were able to seek out the appropriate data sets and then use them within ArcGIS Online to ask and answer geographical questions. By emailing the Electricity Supply Board (ESB) and the City Council asking for data sets, they were also able to map the location of car charging points and recycling centres. They then mapped these with spatial data for income from the Central Statistics Office (CSO). They immediately found that there were bike stands and car charging points in more affluent parts of the city, and more recycling banks and waste sites in more deprived areas of the city. Through using the tool of GIS, the boys were able to identify the outcomes of spatial injustice. To extend students' geographical understanding further and for the students to engage in participatory geography, the next step here would be to consider how future planning decisions might (re)produce or overcome such spatial inequalities.

Overall, the boys were positive about using GIS; their comments included: 'It was good to get to use ArcGIS, as it was very different to anything else we have done in geography'. However, it was not without its challenges: to work so independently was new to the Transition Year students, and at times their emerging expertise on both geography and technology was a challenge. At times, GIS became more of a focus than the enquiries themselves, as it was a novelty for the students to use and there was much to learn. As one boy acknowledged, 'it took a long time to get used to using ArcGIS'. There were also issues with the practicalities of GIS, with one boy stating 'it was annoying when the data sets would not upload', a daily challenge! The experience of success, or enactive attainment according to Bandura (1986), is perceived to be an important source of self-efficacy with GIS (Walshe, 2017). Despites the challenges articulated by boys, their experience of successfully being able to devise and answer their own geographical questions provided a strong foundation for them to continue to develop their use of GIS, especially for the Leaving Certificate Geographical Investigation. The pathway to capitalising upon the value of GIS for geography education for those with little prior experience of GIS appears to involve activities and enquiries that build confidence and understanding around GIS and its functionality (Walshe, 2017; Healy and Walshe, 2020). This appears to be a necessary step for all learners, from primary aged children through to undergraduates and trainee teachers (Ricker and Thatcher, 2017; Walshe, 2017); however, it is possible that securing this foundation earlier might open up new possibilities within geography education as they move to the Senior Cycle in school.

In both of these examples, the learning partially met the criteria for participatory geographical learning, as set out at the start of this chapter. The students had a role in shaping the learning experiences, including the use of GIS, through their questions and ideas that came from their everyday geographies. The students, with the support of their teachers and other adults, collated a range of data through interaction with places to begin to answer their questions. They were engaging with geographical content that transcended their everyday knowledge; for example, through using historical maps to consider how places have changed over time and drawing upon data sets relating to the spatial distribution of facilities across their city. In the second case study, there was greater potential for the young people involved to consider geographical issues that concerned them within their locality. The next step for participatory geographical learning would be to develop opportunities for these students to make informed decisions and contribute their educated perspectives to relevant stakeholders, such as local planners and councillors.

Unlocking the potential of GIS for young people's participatory geography

Young people have always had ideas, opinions and suggestions for their local, national and global communities, both for the present and the future (Pike, 2011; Schlemper et al., 2018; Schlemper et al., 2019). As Hart (1992) argues, propensity to become active citizens needs to be developed through practice at a young age, and this can occur in geography lessons (Pike, 2020). The substantial changes that have occurred in the last decade with how young people interact with their local and wider community through technology (Hammond, 2021) can be harnessed for this process as young people are using spatial digital technologies already. And so enhancing young people's existing digital mapping expertise in school can provide a powerful way to engage young people with the world they live in. Overall geography education can play a critical role in enabling students to understand how 'spatial data is produced, used and shared' (Swords et al., 2019, p.142). Currently in Ireland initiatives such as the Esri ArcGIS Schools Project, which includes partnerships with schools as well as work placements for transition year students, are increasing the take-up of GIS in schools (Esri, 2019). As Biddulph (2018) notes, 'when we know what kind of understanding we wish our students to develop we can plan a coherent curriculum that provides the right opportunities to enable them to do so' (p.101). Moreover, while US research has focussed on the role GIS can play in spatial thinking – 'to ask spatial questions, visualise spatial data and perform spatial analysis' (Butt, 2020, p.96), Fargher (2017b), Walshe (2018) and Healy and Walshe (2020) have placed emphasis on the extent to which GIS can contribute to the geographical knowledge and thinking of students. For this reason, GIS should be considered as a tool to enhance geographical thinking through the development of both geographical knowledge and skills, which in turn can empower students as active citizens. However, there are issues of planning for continuity and progression across primary and secondary schools due to a range of factors, including teacher expertise, expectations of teachers and lack of linkages across phases of education (Catling, 2017). In Ireland this is especially so due to the complicated patterns of feeder schools as well as lack of support for cross phase initiatives. The absence of explicit reference to GIS within official curriculum documents (e.g. DES/NCCA, 2017) might lead to GIS been underutilised and there being a lack of continuity across stages of a student's education in Ireland. However, helpfully for teachers, the newest curriculum documents do contain references to digital spatial technologies (e.g. DES/NCCA, 2018, pp.62–63).

Ultimately, progression in GIS is not merely about students becoming better at using GIS tools, but also requires students to understand how knowledge is constructed within GIS (Fargher, 2017b). Swords et al. (2019) argue that 'GIS can ... serve to disempower ... under-represented social groups and places' (p.142), whilst Elwood (2009) highlights that even within undergraduate geography, there is often a lack of consideration for how GIS are socially and politically constructed. However, a participatory approach to geographical learning, drawing on young people's expertise, appears to provide greater necessity and opportunity for young people to develop this understanding to enable geographically informed 'active citizenship'.

Notes

1 The term 'young people' is used throughout the chapter to refer to anyone under the age of 18, in schools and communities.
2 Children in Ireland start school any time between the ages of 5 and 6, with an optional Transition Year in their 4th year of secondary school; they can be any age between 16 and 19 taking their Leaving Certificate examination.

Acknowledgements

Thank you to the 6th class students of Sacred Heart Boys' National School and the Transition Year students of St Kevin's College for sharing their learning. Special thanks to Jim Ryan, Ian O'Callaghan and Andrew Horan for allowing their classes to share their work for the case studies. Thank you to DCU Geography Specialism students for carrying out fieldwork with the students in Sacred Heart.

References

Alarasi, H., Martinez, J. and Amer, S., 2016. Children's perception of their city centre: A qualitative GIS methodological investigation in a Dutch city. *Children's Geographies*, 14 (4), pp.437–452.

Bandura, A., 1986. *Social Foundations of Thought and Action: A Social Cognitive Theory*. Englewood Cliffs, NJ: Prentice Hall.

Berglund, U. and Nordin, K., 2007. Using GIS to make young people's voices heard in urban planning. *Built Environment*, 33 (4), pp.469–481.

Biddulph, M., 2018. Primary and secondary geography: Common ground and some shared dilemmas. *Teaching Geography*, 43 (3), pp.101–104.

Blatt, A.J., 2015. The benefits and risks of volunteered geographic information. *Journal of Map & Geography Libraries*, 11 (1), pp.99–104.

Butt, G., 2020. *Geography Education Research in the UK: Retrospect and Prospect: The UK Case, Within the Global Context*. London: Springer.

Catling, S., 2003. Curriculum contested: Primary geography and social justice. *Geography*, 88 (3), pp.164–210.

Catling, S., 2017. Everyday geography and pre-service primary geography. In S. Catling (ed.), *Reflections on Primary Geography*. Sheffield: The Register of Research in Primary Geography, pp.178–184.

Chang, C.H. and Kidman, G., 2019. Curriculum, pedagogy and assessment in geographical education: For whom and for what purpose? *International Research in Geographical and Environmental Education*, 28 (1), pp.1–4.

Chawla, L. (ed.), 2002. *Growing up in an Urbanising World*. London: Earthscan.

Danby, S., Davidson, C., Ekberg, S., Breathnach, H. and Thorpe, K., 2016. 'Let's see if you can see me': Making connections with Google Earth in a preschool classroom. *Children's Geographies*, 14 (2), pp.141–157.

Davies, T., Lorne, C. and Sealey-Huggins, L., 2019. Instagram photography and the geography field course: Snapshots from Berlin. *Journal of Geography in Higher Education*, 43 (3), pp.362–383.

Department of Children and Youth Affairs (DCYA), 2015. *National Strategy on Children and Young People's Participation in Decision-Making*. Dublin: Government Publications. www.gov.ie/en/publication/9128db-national-strategy-on-children-and-young-peoples-participation-in-dec/ [Accessed 5 January 2020].

Department of Communications, Energy and Natural Resources (DCENR), 2013. *Doing More with Digital: The National Digital Strategy for Ireland. Phase 1 – Digital Engagement*. Available at: www.dccae.gov.ie/en-ie/communications/topics/Digital-Strategy/Pages/default.aspx?gclid=EAIaIQobChMI1siB2s725gIVBrDtCh1S4ArvEAAYASAAEgJn8vD_BwE [Accessed 8 January 2020].

Department of Education and Science/National Council for Curriculum and Assessment (DES/NCCA), 1999a. *Primary Geography Curriculum*. Available at: www.curriculumonline.ie/getmedia/6e999e7b-556a-4266-9e30-76d98c277436/PSEC03b_Geography_Curriculum.pdf [Accessed 8 January 2020].

Department of Education and Science/National Council for Curriculum and Assessment (DES/NCCA), 1999b. *Primary Geography Introduction*. Available at: www.curriculumonline.ie/getmedia/c4a88a62-7818-4bb2-bb18-4c4ad37bc255/PSEC_Introduction-to-Primary-Curriculum_Eng.pdf [Accessed 8 January 2020].

Department of Education and Science/National Council for Curriculum and Assessment (DES/NCCA), 2003. *Leaving Certificate Geography Syllabus*. Available at: www.education.ie/en/Schools-Colleges/Information/Curriculum-and-Syllabus/Senior-Cycle-/Syllabuses-and-Guidelines/lc_geography_sy.pdf [Accessed 8 January 2020].

Department of Education and Science/National Council for Curriculum and Assessment (DES/NCCA), 2017. *Junior Cycle Geography*. Available at: www.curriculumonline.ie/getmedia/2a7a8d03-00e6-4980-bf20-f58def95688f/JC_Geography-en.pdf [Accessed 8 January 2020].

Department of Education and Science/National Council for Curriculum and Assessment (DES/NCCA), 2018. *Draft Mathematics Curriculum*. Available at: www.ncca.ie/media/3148/primary_mathsspec_en.pdf [Accessed 8 January 2020].

Egiebor, E.E. and Foster, E.J., 2019. Students' perceptions of their engagement using GIS Story Maps. *Journal of Geography*, 118 (2), pp.51–65.

Elwood, S., 2009. Integrating participatory action research and GIS education: Negotiating methodologies, politics and technologies. *Journal of Geography in Higher Education*, 33 (1), pp.51–65.

Esri, 2019. *Lurgan Schools: The Differences we Share*. Available at: https://tinyurl.com/yxg855ap [Accessed 8 January 2020].

Fargher, M., 2017a. GIS and other geospatial technologies. In M. Jones (ed.), *The Handbook of Secondary Geography*. Sheffield: Geographical Association, pp.224–259.

Fargher, M., 2017b. GIS and the power of geographical thinking. In C. Brooks, G. Butt and M. Fargher (eds), *The Power of Geographical Thinking*. London: Springer, pp.151–164.

Fearnley, F., 2021. Social media as a tool for geographers and geography educators. In N. Walshe and G. Healy (eds), *Geography Education in the Digital World*. London: Routledge, pp.75–85.

Giddens, A., 1991. *Modernity and Self-Identity: Self and Society in the Late Modern Age*. Cambridge: Polity Press.

Goodchild, M.F., 1988. Stepping over the line: Technological constraints and the new cartography. *The American Cartographer*, 15 (3), pp.311–319.

Goodchild, M.F., 2007. Citizens as sensors: The world of volunteered geography. *GeoJournal*, 69 (4), pp.211–221.

Hall, T., 2009. The camera never lies? Photographic research methods in human geography. *Journal of Geography in Higher Education*, 33 (3), pp.453–462.

Hall, T., 2015. Reframing photographic research methods in human geography: A long-term reflection. *Journal of Geography in Higher Education*, 39 (3), pp.328–342.

Hammond, L. 2021. Children, childhood and children's geographies: Changing through technology. In N. Walshe and G. Healy (eds), *Geography Education in the Digital World*. London: Routledge, pp.38–49.

Hart, R., 1979. *Children's Experience of Place*. New York: Irvington Publishers.

Hart, R., 1992. *Children's Participation: From Tokenism to Citizenship* (Innocenti Essay No. 4). Florence: International Child Development Centre. Available at: www.unicef-irc.org/publications/100-childrens-participation-from-tokenism-to-citizenship.html [Accessed 11 September 2019].

Healy, G. and Walshe, N., 2020. Real-world geographers and geography students using GIS: Relevance, everyday applications and the development of geographical knowledge. *International Research in Geographical and Environmental Education*, 29 (2), pp.178–196.

Heffron, S.G. and Downs, R.M., 2012. *Geography for Life: National Geography Standards* (2nd edn). Washington, DC: National Council for Geographic Education.

Hicks, D., 2007. Lessons for the future a geographical contribution. *Geography*, 92 (3), pp.179–188.

Hicks, D., 2014. A geography of hope. *Geography*, 99 (1), pp.5–12.

HOSM, 2019. *About Us*. Available at: www.hotosm.org/what-we-do [Accessed 5 September 2019].

Kelly, L., 2018. Geography lesson: Schools map out digital future with tech giant Esri. *Irish Independent*. Available at: www.independent.ie/business/technology/geography-lesson-schools-map-out-digital-future-with-tech-giant-esri-37479626.html [Accessed 5 September 2019].

Kenny, Á., 2019. Armagh students make history by presenting project aimed at tackling sectarianism at world's largest digital mapping conference. *Irish Independent*. Available at: www.independent.ie/irish-news/armagh-students-make-history-by-presenting-project-aimed-at-tackling-sectarianism-at-worlds-largest-digital-mapping-conference-38302786.html [Accessed 5 September 2019].

Laughlin, D.L. and Johnson, L.C., 2011. Defining and exploring public space: perspectives of young people from Regent Park, Toronto. *Children's Geographies*, 9(3–4), pp.439–456.

Lynch, K., 1960. *The Image of the City*. Cambridge, MA: MIT Press.

Martin, F., 1999. Contrasting views on locality between child and adult. *International Research in Geographical and Environmental Education*, 8 (1), pp.78–81.

Massey, D., 2005. *For Space*. London: Sage.

McMahon, G., Percy-Smith, B., Nigel, T., Bečević, Z., Liljeholm Hansson, S. and Forkby, T., 2018. *Young People's Participation: Learning from Action Research in Eight European Cities*. Available at: https://zenodo.org/record/1240227 [Accessed 11 September 2019].

O'Keeffe, G.S. and Clarke-Pearson, K., 2011. The impact of social media on children, adolescents, and families. *Pediatrics*, 127 (4), pp.800–804.

Orben, A., Dienlin, T. and Przybylski, A., 2019. *Social Media's Enduring Effect on Adolescent Life Satisfaction*. Available at: https://osf.io/4xp3v/ [Accessed 9 September 2019].

Pike, S., 2011. 'If you went out it would stick': Irish children's learning in their local environments. *International Research in Geographical and Environmental Education*, 20 (2), pp.139–159.

Pike, S., 2016. *Learning Primary Geography: Ideas and Inspiration from Classrooms*. London: Routledge.

Pike, S. 2020a. Exploring the locality and beyond. *Primary Geography*, 101 (1), pp.20–22.

Pike, S. 2020b. Primary geography for social and environmental justice education. In A.M. Kavanagh and F. Waldron (eds), *Acting for a Better World: Addressing Children's Agency across the Curriculum in the Primary Classroom*. Abingdon: Routledge.

Ramezani, S. and Said, I., 2013. Children's nomination of friendly places in an urban neighbourhood in Shiraz, Iran. *Children's Geographies*, 11 (1), pp.7–27.

Ricker, B. and Thatcher, J., 2017. Evolving technology, shifting expectations: Cultivating pedagogy for a rapidly changing GIS landscape. *Journal of Geography in Higher Education*, 41 (3), pp.368–382.

Roberts, M., 2013. The challenge of enquiry based learning. *Teaching Geography*, 38 (2), pp.50–52.

Sanders, R., 2007. Developing geographers through photography: Enlarging concepts. *Journal of Geography in Higher Education*, 31 (1), pp.181–195.

Schlemper, M.B., Athreya, B., Czajkowski, K., Stewart, V.C. and Shetty, S., 2019. Teaching spatial thinking and geospatial technologies through citizen mapping and problem-based inquiry in Grades 7–12. *Journal of Geography*, 118 (1), pp.21–34.

Schlemper, M.B., Stewart, V.C., Shetty, S. and Czajkowski, K., 2018. Including students' geographies in geography education: Spatial narratives, citizen mapping, and social justice. *Theory & Research in Social Education*, 46 (4), pp.603–641.

Shin, E.E. and Bednarz, S. (eds), 2019. *Spatial Citizenship Education: Citizenship through Geography*. New York: Routledge.

Snapchat, 2018. *Introducing Two New Ways to Reach your Audience by Location and Context*. Available at: https://forbusiness.snapchat.com/blog/location/ [Accessed 3 September 2019].

Soja, E.W., 2009. The city and spatial justice. *Spatial Justice*, 1, pp.1–5.

Swords, J., Jeffries, M., East, H. and Messer, S., 2019. Mapping the city: Participatory mapping with young people. *Geography*, 104 (3), pp.141–147.

Tunstall, S., Tapsell, S. and House, M., 2004. Children's perceptions of river landscapes and play: What children's photographs reveal. *Landscape Research*, 29 (2), pp.181–204.

Turner, A., 2006. *Introduction to Neogeography*. Sebastopol, CA: O'Reilly. Available at: http://site.ebrary.com/id/10762047 [Accessed 5 September 2019].

United Nations (UN), 1989. *Convention of the Rights of the Child*. Available at: www.ohchr.org/en/professionalinterest/pages/crc.aspx [Accessed 8 January 2020].

Walshe, N., 2016. Working with ArcGIS Online Story Maps. *Teaching Geography*, 41 (3), pp.115–117.

Walshe, N., 2017. Developing trainee teacher practice with geographical information systems (GIS). *Journal of Geography in Higher Education*, 41 (4), pp.608–628.

Walshe, N., 2018. Spotlight on … Geographical information systems for school geography. *Geography*, 103 (1), pp.46–49.

Wynne-Jones, S., North, P. and Routledge, P., 2015. Practising participatory geographies: Potentials, problems and politics: Practising participatory geographies. *Area*, 47 (3), pp.218–221.

Young, M.F.D., 2008. *Bringing Knowledge Back In: From Social Constructivism to Social Realism in the Sociology of Education*. London: Routledge.

Part IV

Geographical fieldwork in the digital world

Chapter 12

Using mobile virtual reality to enhance fieldwork experiences in school geography

Rebecca Kitchen

The importance of fieldwork in geography

'Geography wants to take children outside the school and into the streets and fields; it wants to take keyboard tappers out of their gloomy offices and into the rain or the sunshine'. (Bonnett, 2008, p.80). Fieldwork, defined by Lambert and Reiss (2014, p.8) as 'the leaving of the classroom to engage in teaching and learning activities through first-hand experiences of phenomena', has a long and established history in school geography (Cook, 2011). At least part of the reason for this is that it has many conceptual, cognitive, procedural and social benefits which are otherwise difficult to replicate. For example, fieldwork enables students to observe the messy, real world and stimulates their curiosity. It allows them to engage in collaborative and iterative geographical enquiry where they can identify good investigative questions, collect data, analyse this data and draw conclusions (Ofsted, 2008; Lambert and Reiss, 2014; Lambert and Reiss, 2016; Kinder, 2018). It also encourages students to get 'into the streets and fields', which can increase motivation, provide memorable learning experiences and reduce anxiety (Dunphy and Spellman, 2009; Kinder, 2016).

Yet, despite this, the place of fieldwork in geography is arguably 'contested, unclear and under threat' (Lambert and Reiss, 2014, p.8). Rawding (2013) emphasises the centrality of fieldwork to geography, but acknowledges that some perceive that geography can be 'done' (albeit less effectively) without engaging in fieldwork, thereby seeing fieldwork as a desirable but non-essential component of the subject. Coupled with this, there are actual and perceived obstacles which mean that simply taking students out of the classroom can prove challenging (Lambert and Reiss, 2016; Tilling, 2018). The Geographical Association fieldwork survey which had over 250 responses from geography teachers identified the three greatest obstacles as cost, support from the school and teacher planning time (Geographical Association, 2017). Fieldwork is often seen by Senior Leadership Teams as expensive, both in terms of monetary cost – which is usually covered by parental contributions – and curriculum time where geography competes with other subjects. It also has its own distinctive pedagogies. In addition, teachers – given that most fieldwork is led by teachers – need support, training and time for planning in order for the fieldwork to be well-organised and effective.

The introduction of a compulsory fieldwork component in both the reformed GCSE and A-Level geography specifications in England has served to move the discussion regarding effective fieldwork to the fore (Department for Education, 2014a; Department for Education, 2014b). As Winter (2017) highlights, teaching priorities are fundamentally determined by the prescribed curriculum content, performance in high-stakes exams (such as GCSE and A-Level) and the subsequent public scrutiny of the result. Therefore, Tilling (2018) cites this, coupled with the strong culture and tradition of fieldwork amongst geography teachers, as contributing to the increase between 2005 and 2017 in geography fieldwork teaching days of 23.5% at Field Studies Council centres.

Given this backdrop of the central role of fieldwork in geography, whilst also being mindful of the logistical obstacles that can be presented for teachers, it is important that fieldwork experiences are included and are effective in making best use of the time spent in the field. Technology-based virtual fieldwork experiences which utilise videos, web-based interactive experiments or simulations have been suggested as a way to improve the effectiveness of physical fieldwork[1] (Robinson, 2009; Stoddard, 2009; Argles et al., 2015). Despite this, the literature surrounding the use of virtual reality (VR) in enhancing physical fieldwork is sparse (Maskall and Stokes, 2008; Stoddard, 2009). Therefore, this chapter seeks to begin address this. VR and Google Expeditions (GE) are first defined, followed by a description of how they can be used before, during and after physical fieldwork in order to enhance its effectiveness. Finally, issues and implications for practice will be considered.

Virtual reality

The term 'VR' was coined in the 1980s. However, the technologies used to simulate both realistic and non-realistic environments have their foundations dating back to the development of the stereoscope in 1838 (Franklin Institute, 2019). The characteristics of VR include the combination of 3D images with sounds, sensations and features that enable user interactivity with the simulated environment. These environments can be accessed through technologies which range significantly both in terms of cost and mobility, factors that are significant for teachers wishing to engage in VR fieldwork (Minocha and Tudor, 2017). For example, Second Life – an online virtual world where avatars (virtual representations of the user) socialise, engage in activities, create and trade – can be accessed on desktops without the use of a VR headset (Rymaszewski et al., 2007; Linden Labs, 2019). At the other end of the mobility spectrum is Google Cardboard where users wear a low-cost (typically between £5 and £10) VR headset and smartphone combination to access 360-degree photospheres (Google VR, 2019).

Google Expeditions

GE is an app (available on Android and iOS platforms) which contains a large and growing library of over 1000 'expeditions'. These expeditions typically fall into three categories – physical locations, simulations (e.g. the process of photosynthesis or pollination) and career expeditions (e.g. a day in the life of a software developer), of which physical locations have the greatest relevance in the context of geography fieldwork. Each expedition is composed of one or more photospheres which enable the visualisation of locations such as the Great Barrier Reef or London Olympic Park which may be difficult to visit in real life (Minocha et al., 2017). Using the GE app on a tablet, the 'guide' (usually the teacher) uses the explanatory text, points of interest and questions within the expedition to lead students around the environment. The students use the GE app in 'follower' mode and view the environment through the app on a smartphone which is placed in the Google Cardboard headset.

Using Google Expeditions to enhance physical fieldwork

In attempting to draw out how VR, and specifically GE, may be used to enhance physical fieldwork, this chapter describes a year-long research project funded by Google's Virtual

Reality Research Programme and carried out by the Open University in collaboration with the Field Studies Council (FSC), Association for Science Education (ASE) and Geographical Association (GA). The research focus was to investigate how mobile VR could be integrated into science and geography fieldwork – crucially not as a replacement for physical fieldwork but as an enhancement. The data collection involved semi-structured interviews with six curriculum experts and 20 teachers (nine of whom were geography specialists) and observations of 24 classroom-based lessons which involved a total of 549 students (255 in geography) in Years 4–11. In addition, five workshops were held; four with 55 geography and science teachers, and a separate workshop was held for 19 FSC staff. The researchers also attended one physical fieldtrip which involved 68 Year 7 students and their three geography teachers visiting a local nature reserve in the Chilterns, South East England (Minocha et al., 2018; Tudor et al., 2018). Using the data, particularly the interviews that came out of the teacher and FSC staff workshops, the researchers developed suggestions of how virtual realities may enhance physical fieldwork. It is important to note that these suggestions are based on the participant's perceptions and are referred to in the literature as 'affordances' – i.e. the perceived and actual properties of an object, in this case GE, that determine how the object could possibly be used (Soegaard, 2017; Tudor et al., 2018).

Maskall and Stokes (2008) identify three phases of fieldwork. Pre-fieldwork induction, which includes the development of key enquiry questions for investigation, the actual physical fieldwork experience, which involves some form of data collection, and a post-fieldwork reflective phase with a focus on analysis and evaluation. Coupled with this, Remmen and Frøyland (2013; 2014; 2015) have suggested that the impact of fieldwork is increased if each aspect of this three-part structure is engaged with effectively. Consequently, this structure is used here to identify the ways in which GE can be used to enhance physical fieldwork at each stage.

Pre-physical fieldwork phase

Given the pressures on time spent in the field identified earlier in this chapter, it is important that the pre-physical fieldwork phase provides students with the tools and skills for observation and investigation that they will need in the field, as well as familiarising them with the environment so as to make the most effective use of their time within it. Participants suggested that VR more generally, and GE specifically, could be used to achieve these aims in the ways discussed in the following sections.

Understanding fieldwork in geography

There are many stakeholders involved in physical fieldwork. In addition to the geography teachers and students, there are their parents, members of the Senior Leadership Team and other non-specialist teachers and support staff who frequently attend physical fieldwork but may not have any experience or training in this area. Providing experience of an environment within GE for some of these stakeholders, even if it is not the same as the location planned for the fieldwork, can have multiple benefits. It can encourage them to think about the distinctive nature of fieldwork and what might be involved. In addition, by engaging with similar virtual environments, non-specialists can be trained to support the physical fieldwork, to ask effective questions and to subsequently focus their planning. For example, it allows them: 'to see how accessible it [the location] is, is there a road network close by? How long it takes them to get there. How long is it going to take to actually access the site they like' (FSC staff in Minocha et al. (2018, pp.8–9)).

Familiarisation with the locations of the fieldwork

Experiencing the environment that they are going to visit in virtual reality can bring a familiarity which can support both conceptual understanding and also logistical planning: 'for timing … this will allow them [students] to plan ahead for how long it will take them to access the site and to carry out the physical measurements' (FSC staff in Minocha et al. (2018, p.8)). McDougall (2019) does warn that such preparation can reduce the initial impact of being physically in the environment for the first time – the so-called 'wow factor' – which students often experience. However, he does concede that familiarity does enable students to embark on their physical fieldwork with a better understanding of both the topic and location and this increases focus. Teachers should also be aware that often the 'wow factor' is not lost completely but is merely displaced to the point at which they engage with the environment through the technology (Parkinson, 2017).

It is worth noting that the Google Cardboard app (available on both Android and iOS) can be used to take 360-degree photospheres which can then be used within GE (Alba, 2015). Therefore, teachers can create bespoke expeditions of the locations that students will visit and direct them to specific points of interest. This also provides the opportunity for students to experience the environment during a different season, for example, so that they can identify temporal variation.

Comparing and contrasting locations and habitats

Whilst no two landscapes are exactly alike, GE can be used prior to visiting a landscape to explore a range of similar and different environments. These are typical activities used in geography education and serve to build the confidence of students in landscape and process interpretation prior to their visit. For example, McDougall (2019) suggests that prior to physical fieldwork in a deglaciated landscape such as the Lake District or Snowdonia, students may engage in a virtual visit to a contemporary glaciated environment, e.g. Mont Blanc, in order to support their sense-making.

Risk assessment

As Cook (2012) identifies, there is a lack of consensus surrounding the concept of risk within the context of physical fieldwork; some teachers place the onus of negotiating risk firmly with the student whilst others take full responsibility themselves. She further notes that fieldwork can provide a valuable opportunity for students to experience the 'dignity of risk'; this is the idea that self-determination and the right to take reasonable risks are essential for dignity and self-esteem (Wolpert, 1980). However, given the diversity of positions regarding risk and recognising that some teachers are over-cautious and concerned about their duty of care, this can mean that some students are denied this opportunity. It is therefore vital that teachers engage with their students' perceptions of risk, and this can be achieved prior to going into the field by discussing potential risks within different GE environments, which is highlighted by a geography curriculum expert from the GA:

> you can go out in the field and talk about risk assessment but you need to have done that beforehand. The students need to be good at looking at what the risks might be. Showing it to them beforehand and going 'Okay, what do you think the risks are going to be?' would be really useful. (Minocha et al., 2018, p.8)

Observation skills

Haynes (2007), in her work with children in Early Years, asserts that in order to develop an appreciation for the world around them children must learn to observe and talk about their observations. Familiarity with an environment coupled with practice in observation is likely to enable students to ask more effective questions and plan their data collection which they then apply in the field. An FSC environmental science expert outlines the unfamiliarity that they are often faced with, as 'An awful lot of people go on trips, not only they haven't been to that site, they haven't been to anything that looks remotely like that site, sometimes never before in their lives' (Minocha et al., 2018, p.9).

It is also worth highlighting that, from my experience, the ability to observe within GE is heightened compared to traditional 2D photographs. The quality of the photospheres allows detail to be visible, and the wearing of the headset and immersive nature of the experience enables the blocking of external distractions and thus a focus on what can be observed within the virtual environment. For example, when observing within the 'Environmental Change in Borneo' expedition, droplets of rainwater were visible on the leaves in the shrub layer which drew my attention and allowed me to develop questions based on that specific phenomenon rather than the environment more generally.

Developing enquiry questions for investigation

Students do not necessarily come into the classroom (or, for that matter, the field) wanting to know about the concept that they are about to study (Roberts, 2013). Yet, if they are to learn about this concept they need to have some interest and engage at some level with the ideas that are being introduced. In order to spark curiosity, or as Roberts terms it a 'need to know', stimulus material such as a GE expedition can be used to connect new concepts with students' known experience. The fact that the virtual experience is single-user, i.e. the student does not interact with others within the environment, also means that it is relatively personalised, which can also increase their curiosity. A science curriculum expert from the ASE suggests:

> if you were doing it [the lesson] in virtual reality, students could come up with the investigative questions, think what approach they would use and even start trying out the various ways of approaching the question, without any outside influence from the teacher. (Minocha et al., 2018, p.7)

As students' curiosity is sparked, they are more likely to generate specific questions about the environment. Minocha et al. (2018) highlight that teachers in their study, where GE was used to generate enquiry questions with students, reported a greater prevalence of higher-order questions than more traditional stimuli such as videos, photos or text. It also provides an opportunity for students to pilot their questions and potential data collection methods before arriving in the field, as suggested by an FSC fieldwork and field sciences expert:

> it [the lesson incorporating VR] would be like a pilot study ... The big abilities [via GEs] would be to get familiar with the location, ask questions about that location and plan methods that work in the context of that place instead of [deriving] quite generic questions. (Minocha et al., 2018, p.7)

Physical fieldwork phase

One might assume that because of the limitations of the technology and the importance of focusing on the real world during physical fieldwork, VR may not have a place at this stage. However, Tudor et al. (2018) suggest that rather than being a distraction, GE can provide and articulate different perspectives in the field. They describe Year 7 fieldwork in the Chilterns where students were investigating the potential impact of the development of HS2[2] on a local nature reserve. During the fieldwork, students had the opportunity to investigate the 'Environmental Change in Borneo' expedition to help them to understand the impact of deforestation, land clearance and the development of buildings in a distant place, in order to sensitise them to the potential impact of similar activities in their local area. Using VR in the field can therefore enable students to move from an international to a local context, which in turn can develop their awareness, knowledge and understanding, and attitudes and values.

Awareness

Two types of awareness were identified by Tudor et al. (2018): awareness of environmental challenges and awareness of learning about the characteristics of places, which are demonstrated by this comment from a Year 7 student:

> it helped me [to understand] because I saw a beautiful jungle full of green and life. Then in virtual form we saw a building site, sparse trees. If that happened to the Chilterns, with for example HS2, it would be devastating. (Tudor et al., 2018, p.32)

Whilst students may have a superficial or theoretical awareness of both of these aspects prior to going into the field, the combination of GE and being in the environment can enhance awareness and make students more cognisant of its impact. Specifically, it was achieved by enabling the comparison between a distant place and the location being studied. Also, being guided by the teacher within the expedition emphasised particular features and raised student awareness of them.

Knowledge and understanding

Students' knowledge and understanding regarding the geographical concept of scale was also developed. The VR enabled them to have a 3D bird's eye view over the rainforest, and this gave them both a different perspective and an idea of the vastness of the impact of deforestation. Coupled with this, using VR in the field encouraged students to compare and contrast what was happening in their local nature reserve in the Chilterns with the Borneo rainforest; it took the global to the local (and vice versa). They could infer the broader impact of human-induced change on the environment and could discuss how the construction of HS2 might affect biodiversity, tourism and the local community: 'It made me understand that there would be a big change in the Chilterns and not necessarily a good change. Also, it will ruin it for the wildlife and animals which help the Chilterns grow and expand' (Year 7 student in Tudor et al. (2018, p.32)).

Attitudes and values

Students' attitudes towards the environmental changes were dominated by negative statements. They described the impacts on the Borneo rainforest as being worse than they had expected and felt that this perception of space and sense of scale was enhanced by the VR beyond that which could be provided in a 2D photograph. Beyond this, using VR in the field also appears to enable students to view beyond what is immediately visible and to access locations which are otherwise inaccessible: 'GE gives you a different perspective, the first person view, a kind of outsider view that you wouldn't get without a helicopter or an aerial view' (FSC fieldworker and field sciences expert in Minocha et al. (2018)). This extends to students with disabilities, including hidden disabilities such as anxiety, who may be unable to access fieldwork locations or fully participate in the activities. However, whilst McDougall (2019) suggests that this therefore can increase inclusivity, Healey et al. (2002) disagree. They argue that although virtual reality may bring new learning experiences for disabled students, this sidesteps the main issue of access to the curriculum. They therefore believe that engaging in virtual reality for this purpose only actually serves to exclude students, and that alternative solutions, such as adjusting the objectives and outcomes of the fieldwork, are more inclusive.

Post-physical fieldwork phase

The post fieldwork phase allows opportunities for de-briefing and reflection. If students are able to re-visit the location virtually then this can prompt memories of the physical fieldwork and enable them to stand back and look critically at what they did.

Evaluation

One of the aspects of the reformed GCSE fieldwork that students find challenging is the need to evaluate what they have done in the field at each stage of the enquiry process (FSC, 2019). VR enables this reflection so that rather than prompt students to recall w*hat* they did, they can understand w*hy* they did it, as explicated by a geography teacher here:

> When they come back from a field trip, they've got to be able to understand how reliable [is] the [work] they did? So that will certainly be a good way. If they had an expedition, they could really look at the flaws in what they did and did they choose the right number of sample sizes? (Minocha et al., 2018, p.10)

Applying their knowledge to different locations

Students also struggle to apply their fieldwork knowledge to new and different locations, which is another feature of the reformed GCSE specifications (FSC, 2019). Typically, questions present an unfamiliar fieldwork scenario which students have to comment on, and this can relate to any aspect of the enquiry process. However, if students do not have experience of the type or landscape of fieldwork presented or the techniques used, then they can lack the confidence to answer. Virtual fieldwork provides the opportunity for students to practice techniques and apply their skills in different contexts and locations, elaborated on here by a geology educator: 'Once they've analysed their (field) data, they've realised they could have done it somewhere else. Then they could go and do it virtually somewhere else' (Minocha et al., 2018, p.10). Extending this point, an FSC fieldwork and field science expert proposes that students can '… try to transfer the skills that they've developed in a physical field trip into … the skills they know by using them in a different setting and that different setting is a virtual one' (Minocha et al., 2018, p.10).

Issues and implications for practice

The potential for VR applications, such as GE, to enhance outdoor experiences at various stages of the process has been outlined, and this area is certainly a fertile ground for further research. However, a chapter such as this would not be complete without a consideration of the issues and implications for practice. After all, it doesn't matter how effective the technology is if the barriers to using it are, or are perceived to be, too great. The greatest concerns for teachers are likely to surround the practical implications of using the technology; the cost of setting up the equipment, whether it is practical to implement both in the classroom and in the field, and their own confidence in using it.

As Parkinson (2017) identifies, for most schools, using GE would require a fairly substantial investment and it is likely that its use would need to be justified across the curriculum. Assuming that single Google cardboard headsets can be purchased for around £10 each, an outlay of around £100 would be needed to enable small group work in GE. The headsets are relatively robust but ultimately they are made of cardboard and so may not withstand much heavy handling or regular use. In addition, each headset also requires a smartphone to run the app, and this would either require a more substantial investment or a robust mobile phone policy which allows students to use their own smartphones in school (Welsh and France, 2012).

As with any use of technology, teachers also need to feel comfortable with both the working of the technology and the pedagogical considerations that arise from using it. The tours need to be downloaded, which requires time and a reliable WiFi network. It is also worth highlighting that some students are susceptible to dizziness, eye-strain and headaches whilst using VR equipment, and so care needs to be taken and the appropriate warnings issued before engaging with the activities (Parkinson, 2017; Science Focus, 2019). These practical considerations are exacerbated in the field where inclement weather may mean that the devices cannot be used. Even if the weather is fine, the smartphones and tablet need to be charged for the duration of the trip and there is a reliance on a battery-powered router to connect the devices (Parkinson, 2017; Tudor et al., 2018).

Conclusion

Technology should never be used as a gimmick, either in the classroom or the field, and it certainly must never replace physical fieldwork; children need to experience the real world in order to make sense of it (Bonnett, 2008). However, this chapter has described how VR, and GE in particular, can be used to enhance the geography that students do in the field by turning it from a one-off experience into an experience that is embedded within their geographical learning with pre-fieldwork preparation and follow-up; this wraparound approach means there is greater opportunity for fieldwork to be carefully planning and reflected on by both students and teachers. It scaffolds a deeper learning that is difficult to re-create in other, more traditional, ways and offers a personalised experience that can be more relatable to the student.

Certainly, there are obstacles to implementing VR both in the classroom and in the field, but as the technology develops and becomes more affordable these are likely to become negligible. Coupled with this, there are now opportunities for teachers to develop supporting curriculum materials and their own photosphere content which has the potential to broaden the expeditions available and deepen the learning experience even further.

Notes

1 It is important to note that, in this context, the term physical fieldwork refers to the student physically going into the field to carry out fieldwork, rather than the student engaging in fieldwork pertinent to physical geography.
2 HS2 or High Speed 2 is a rail network that will connect London, Manchester, Birmingham and Leeds with over 345 miles of high-speed track (HS2, 2019).

References

Alba, D., 2015. *Google's Cardboard Camera App Makes Anyone a VR Photographer*. Available at: www.wired.com/2015/12/google-cardboard-camera-app/# [Accessed 20 December 2019].

Argles, T., Minocha, S. and Burden, D., 2015. Virtual field teaching has evolved: Benefits of a 3D gaming environment, *Geology Today*, 31 (6), pp.222–226.

Bonnett, A., (2008) *What is Geography?* London: Sage.

Cook, V., 2011. The origins and development of geography fieldwork in British Schools. *Geography*, 96 (2), pp.69–74.

Cook, V., 2012. Conceptualising 'risk', *Teaching Geography*, 37 (2), pp.54–56.

Department for Education, 2014a. *Geography GCSE Subject Content*. London: Department for Education. Available at: https://assets.publishing.service.gov.uk/government/uploads/system/uploads/attachment_data/file/301253/GCSE_geography.pdf [Accessed 26 December 2019].

Department for Education, 2014b. *GCE, AS and A Level Subject Content for Geography*. London: Department for Education. Available at: https://assets.publishing.service.gov.uk/government/uploads/system/uploads/attachment_data/file/388857/GCE_AS_and_A_level_subject_content_for_geography.pdf [Accessed 26 December 2019].

Dunphy, A. and Spellman, G., 2009. Geography fieldwork, fieldwork value and learning styles. *International Research in Geographical and Environmental Education*, 18 (1), pp.19–28.

Field Studies Council, 2019. *FSC Making your GCSE Fieldwork Exam Ready* [webinar, April/May]. FSC and The Geographical Association.

Franklin Institute, 2019. *History of Virtual Reality*. Available at: www.fi.edu/virtual-reality/history-of-virtual-reality [Accessed 24 December 2019].

Geographical Association, 2017. Findings from the 2016 Geographical Association fieldwork survey. Paper presented at the Association for Science Education Annual Conference, Birmingham, UK, 6 January.

Google VR, 2019. *Google Cardboard*. Available at: https://arvr.google.com/intl/en_uk/cardboard/ [Accessed 23 December 2019].

Haynes, S., 2007. The foundations of fieldwork. *Primary Geographer*, 63 (2), pp.18–19.

Healey, M., Roberts, C., Jenkins, A. and Leach, J., 2002. Disabled students and fieldwork: Towards inclusivity? *Planet*, 6 (1), pp.24–26. Available at: www.tandfonline.com/doi/full/10.11120/plan.2002.00060024 [Accessed 20 December 2019].

HS2, 2019. *What is HS2?* Available at: www.hs2.org.uk/what-is-hs2/ [Accessed 26 December 2019].

Kinder, A., 2016. The value of fieldwork. *GA Magazine*, 32, p.19.

Kinder, A., 2018. Acquiring geographical knowledge and understanding through fieldwork. *Teaching Geography*, 43 (3), pp.109–112.

Lambert, D. and Reiss, M., 2014. *The Place of Fieldwork in Geography and Science Qualifications.* London: Institute of Education, University of London.

Lambert, D. and Reiss, M., 2016. The place of fieldwork in geography qualifications, *Geography*, 101 (1), pp.28–34.

Linden Labs, 2019. *Second Life.* Available at: https://secondlife.com/ [Accessed 23 December 2019].

Maskall, J. and Stokes, A., 2008. *Designing Effective Fieldwork for the Environmental and Natural Sciences.* Available at: www.advance-he.ac.uk/knowledge-hub/designing-effective-fieldwork-environmental-and-natural-sciences [Accessed 26 December 2019].

McDougall, D., 2019. Spotlight on …VR glaciers and glaciated landscapes. *Geography*, 104 (3), pp.148–153.

Minocha, S. and Tudor, A., 2017. *Virtual Reality in Education – Learning Design and Technology Enhanced Learning Special Interest Group.* Milton Keynes: The Open University. Available at: http://oro.open.ac.uk/51326/ [Accessed 23 December 2019].

Minocha, S., Tilling, S and Tudor, A., 2018. *Role of Virtual Reality in Geography and Science Fieldwork Education* (Knowledge Exchange Seminar Series: Learning from New Technology). Available at: http://oro.open.ac.uk/55876/ [Accessed 26 December 2019].

Minocha, S., Tudor, A., Tilling, S. and Kitchen, R., 2017. Investigating the role of virtual reality in geography via Google Expeditions. Paper presented at the Geographical Association Annual Conference. Available at: www.geography.org.uk/write/MediaUploads/Training%20and%20events/GAConf17_Workshop2.pdf [Accessed 22 December 2019].

Ofsted, 2008. *Geography in Schools: Changing Practice* (Ofsted Ref 070044). Available at: www.geography.org.uk/download/ofsted%20report%20good%20practice%20in%20schools%20-%20changing%20practice%202008.pdf [Accessed 26 December 2019].

Parkinson, A., 2017. Is VR the real deal? *Independent Education Today.* Available at: https://ie-today.co.uk/Blog/is-vr-the-real-deal/ [Accessed 20 December 2019].

Rawding, C., 2013. *Effective Innovation in the Secondary Geography Curriculum: A Practical Guide.* Abingdon: Routledge.

Remmen, K. and Frøyland, M., 2013. How can students be supported to apply geoscientific knowledge learned in the classroom to phenomena in the field: An example from high school students in Norway. *Journal of Geoscience Education*, 61, pp.437–452.

Remmen, K. and Frøyland, M., 2014. Implementation of guidelines for effective fieldwork designs: Exploring learning activities, learning processes, and student engagement in the classroom and the field. *International Research in Geographical and Environmental Education*, 23 (2), pp.103–125.

Remmen, K. and Frøyland, M., 2015. What happens in classrooms after earth science fieldwork? Supporting student learning processes during follow-up activities. *International Research in Geographical and Environmental Education*, 24, pp.24–42.

Roberts, M., 2013. *Geography through Enquiry.* Sheffield: Geographical Association.

Robinson, L. 2009. Virtual field trips: The pros and cons of an educational innovation. *Computers in New Zealand Schools: Learning, Teaching, Technology*, 21 (1), pp.1–17.

Rymaszewski, M., Wagner J., Wallace, M., Winters, C., Ondrejka, C. and Batstone-Cunningham, B., 2007. *Second Life: The Official Guide.* Indianapolis, IN: Wiley.

Science Focus, 2019. *Are VR Headsets Bad for Your Health?* Available at: www.sciencefocus.com/future-technology/are-vr-headsets-bad-for-your-health/ [Accessed 26 December 2019].

Soegaard, M., 2017. *Affordances: The Glossary of Human Computer Interaction.* Available at: www.interaction-design.org/literature/book/the-glossary-of-human-computer-interaction/affordances [Accessed 26 December 2019].

Stoddard, J., 2009. Toward a virtual field trip model for the social studies. *Contemporary Issues in Technology and Teacher Education*, 9 (4), pp.412–438.

Tilling, S., 2018. Ecological science fieldwork and secondary school biology in England: Does a more secure future lie in Geography? *The Curriculum Journal*, 29 (4), pp.538–556.

Tudor, A., Minocha, S., Collins, M. and Tilling, S., 2018. Mobile virtual reality for environmental education. *Journal of Virtual Studies*, 92 (2), pp.25–36.

Welsh, K. and France, D., 2012. Spotlight on … Smartphones and fieldwork. *Geography*, 97 (1), pp.47–51.

Winter, C. 2017. Curriculum policy reform in an era of technical accountability: 'Fixing' curriculum, teachers and students in English schools. *Journal of Curriculum Studies*, 49 (1), pp.55–74.

Wolpert, J., 1980. The dignity of risk. *Transactions of the Institute of British Geographers*, 5 (4), pp.391–401.

Chapter 13

Teaching and learning geography with mobile technologies and fieldwork

Chew-Hung Chang

Introduction

There is little doubt that 'geographical education is indispensable to the development of responsible and active citizens' (International Geographical Union – Commission on Geographical Education (IGU-CGE), 2016, p.1). This is of particular importance as we are faced with unprecedented environmental and social change (Morgan, 2019). Undeniably, school geography 'offers the opportunity [for students] to acquire knowledge and skills to see … what we can do differently' (Béneker and van der Schee, 2015, p.287), which emphasises the future-orientated nature of the subject for our students (Lambert, 2016). This outlook complements the aims of geographical education as outlined in the International Charter on Geographical Education (IGU-CGE, 1992; 2016) where the purpose of geographical education is to allow students to 'face questions of what it means to live sustainably in this world … [and] … how to exist harmoniously with all living species' (IGU-CGE, 2016, p.3). The assertions on the importance of helping children to live sustainability and to exist harmoniously with all living beings are aligned with the United Nations Educational, Scientific and Cultural Organisation (UNESCO) Delors (1996) report, which included four pillars as key concepts to developing education for the 21st century:

1. Learning to know – having broad general knowledge and also depth in a few subjects.
2. Learning to do – to acquire vocational/occupational skills as well as the capability to respond to many situations.
3. Learning to be – to develop one's personality and to be able to act with growing autonomy, judgment and personal responsibility.
4. Learning to live together – to develop an understanding of other people and an appreciation of interdependence.(Delors, 1996, p.37)

Apart from the unparalleled changes to the physical environment, students are also growing up in a digital world of rapidly developing technologies (Hammond, 2021). As information is becoming readily available at the click of a button, there is an impression that knowledge has become ubiquitous. The argument that one can look for almost any information on the internet is flawed, especially when considering how one must know what to key into the search field in the first place. How does one search for that which one does not know? Having the wisdom to know what to search for and not just blindly searching for facts will be a hallmark of learning in the digital age.

Another problem associated with a generation of students that is so familiar with getting information from mobile technologies is the distancing of students from the real-world context. Their social experiences and interactions are built on digital representations of these interactions. How can children learn about the world if they are only experiencing it vicariously through virtual simulations on their devices? But learning in the real world and using mobile technologies need not be mutually exclusive and can be used together to achieve better learning outcomes. We need to move beyond learning to know to learning to be. The use of fieldwork and technologies for teaching and learning geography has been established over the last two decades (Gerber and Goh, 2000; Hsu and Chen, 2010; Medzini et al., 2015; Jarvis et al., 2016; Carbonell Carrera et al., 2018). For example, use of fieldwork and technology have each been highlighted for their importance to geographical education in the International Charter on Geographical Education since 1992 (IGU-CGE, 1992).

Trends in research on mobile technologies and fieldwork in geography education

In a survey of all articles published in the journal *International Research in Geographical and Environmental Education (IRGEE)* and the *Journal of Geography (JoG)* from 2009 to 2018 (inclusive), a total of 23 articles were about the use of mobile technologies and fieldwork in teaching and learning geography (11 from *IRGEE* and 12 from *JoG: Annex A*). The methodology employed was to examine the articles at the title level for reference to mobile technologies or field studies or fieldwork. These two geography education journals were selected because they are seminal journals in international geography education: *IRGEE* was started in 1992 at about the same time as the 1992 Charter's call for a more concerted effort in research on geographical and environmental education. *IRGEE* is also the flagship journal of the IGU-CGE, while *JoG* is the flagship journal of the National Council for Geographic Education (NCGE) based in the United States. Both journals are the key publication outlets for research on geography education by scholars in the field. The journals are also the highest cited in the field, with H-indices of 19 and 25 for *IRGEE* and *JoG* respectively. The titles of the 23 articles were then coded using an open coding scheme which resulted in eight themes (Table 13.1). These theme categories are not mutually exclusive, as some articles related to more than one theme.

The majority of the studies were about fieldwork (20 out of 23), with much fewer about mobile technologies (8 out of 23). Laws' (1989) assertion that the discipline of geography is

Table 13.1 Frequency of occurrence of words within article titles about mobile technology and fieldwork in *International Research in Geographical and Environmental Education* and *Journal of Geography*.

Theme/keyword	Frequency
Field	20
Learning	9
Trips	5
Skills/using	5
Mobile	4
Experiences	4
Course	3
Benefits	2

highly dependent on good fieldwork could explain why there is greater reference to fieldwork in the research literature. The use of fieldwork in geography has been a signature practice that was well-established before the use of mobile technologies. Mobile technologies with their current state of affordances have been a recent phenomenon in the last decade (Hsu and Chen, 2010; Carbonell Carrera et al., 2018), and so this might explain why there have been fewer articles in this area. These articles are about the use of GIS and internet affordances that could technically be accessed using mobile devices. However, focusing only on the title means that it is not possible to identify whether mobile technologies were explicitly referenced within the article.

A further analysis in the form of a word cloud (Figure 13.1) was derived from the abstracts of the articles cited in Annex A. This illustrates a clear focus on the student and the field learning context. Words like 'data' and 'field' are also commonly used in the abstracts, indicating that the learning occurs through the collection and utilisation of data from the field in supporting the students' learning.

Mobile technologies and geography education

Technology can help enhance students' ability to think spatially with geospatial technologies such as GIS (Bednarz and Bednarz, 2004). In particular, mobile technology has the potential to take the spatial thinking a step further by immersing students in the real world, taking the student beyond learning to know to learning to do and learning to be (e.g. Healy and Walshe, 2020; Pike, 2021). Although technological innovations have significantly improved visualisation of space through three-dimensional (3D) representations, such as virtual reality (VR) and augmented reality (AR), this type of simulation of the physical world only provides a proxy experience of the actual dynamics of spatial variation (see Kitchen (2021) and Priestnall (2021) for further exploration of VR and AR opportunities respectively). If

Figure 13.1 Word cloud showing occurrence of words about mobile technology and fieldwork within abstracts from articles in Annex A (articles from *International Research in Geographical and Environmental Education* and *Journal of Geography*).

students are able to use these visualisation technologies on their mobile devices, then the real world can be set as a backdrop on which computer-generated content can enhance learning (Yuen et al., 2011; Priestnall, 2021; Schaal, 2021).

I argue that there is no escaping the pace of the accelerated exchange of information between people through the internet and mobile technologies. Across web applications and different platforms, people are able to access and update information within microseconds. The advancement of web and mobile technologies has allowed developers to implement functions originally developed for desktop personal computer applications on mobile devices. These innovations allow a move away from expert-centred information flow to a new way of gathering data from collective intelligence and turning the web into a kind of global brain (O'Reilly, 2007) through mobile devices. For schools, students can upload data to the server and share data with other students. In particular, students can act as citizen sensors to share volunteered geographical information (VGI; Goodchild, 2007) to mobile GIS platforms. One successful application for users to share reviews about local businesses such as eateries, which can in turn be used by other users to find information such as rating and price range, is Google Maps. Users can even pin the location to share with others. Another example is the Waze mobile application which is a community-based traffic and navigation platform that shares real-time traffic information by amassing traffic data from volunteer drivers. For instance, if there is an accident, a Waze user can share the location and timing of the accident on-site, and other users can be alerted to divert their route or adjust their expected arrival time. Other users can also verify this information through the app as they pass by the reported location. These applications utilise crowd-sourced content, and can potentially help geographical learning as students can examine their own spatial understanding. This will take the student beyond learning to do to learning to live together as they collectively build an understanding of the geographical phenomena they encounter. Wu (2013) examined students' understanding of the world and whether they were able to use spatial terms (e.g. countries, concepts of space, direction and topology) appropriately to describe geographical phenomena through students' VGI data. In this study, Wu (2013) proposes that the use of technology enables teachers and students to evaluate their spatial understanding, which emphasises how technology can be used to establish the next steps for students' geographical learning. However, as Roberts (2021) highlights, the internet means that both geography teachers and students can be inundated with information, and this is why it is ever more important that geographical learning using mobile technologies enables students to engage with the information they have sought, so that students can critically 'employ the disciplinary knowledge of geographical thinking to explain, analyse, evaluate, form an opinion and maybe even take action on what they have learnt' (Chang and Wu, 2018, p.36).

Teacher engagement with technologies: Technological pedagogical content knowledge

As teachers gain teaching experience and expertise through initial teacher education and professional development, they develop pedagogical content knowledge (PCK), which enables them to sequence lessons, design activities, choose resources and craft assessment tasks, and ensure high-quality learning for their students (Shulman, 1986). By extension the technological pedagogical content knowledge (TPACK) of the teacher is paramount in the successful design and implementation of geographical learning with mobile technologies

(Chang and Wu, 2018). The discussion is focused on technologies in general, but is applicable to the context of mobile technologies as TPACK provides a theory and framework (see Figure 13.2) to capture the knowledge teachers need to teach a subject with use of technology (Mishra and Koehler, 2006).

At the centre of the TPACK model is the intersection of content (CK), pedagogy (PK), and technological (TK) knowledge. While technological knowledge about affordances and constraints can help design effective learning activities, a good geographical learning activity must be guided by the best pedagogical strategy for a specific topic, theme or concept. The intersection between the domains of knowledge and how a teacher can use technology in designing teaching is affected by the interplay between their TK, PK and CK. Issues such as teacher quality, student profile, school culture, resource endowment and the topic to be taught will affect a teachers' TPACK. Chang and Wu (2018) argue that TPACK is an

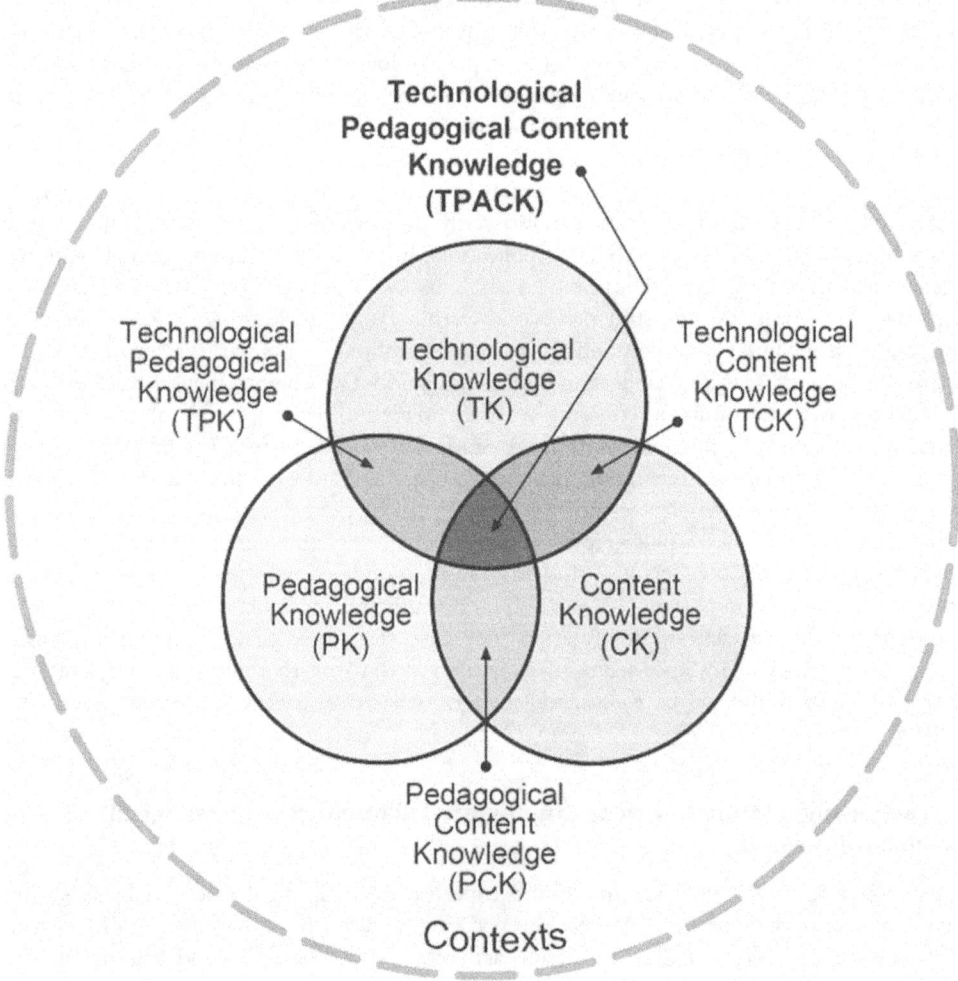

Figure 13.2 The TPACK framework.
Source: Reproduced by permission of the publisher, © 2012, tpack.org.

operationalisation of the interaction between the teacher, the subject matter (school geography in this case) and the student's experiences. Instead of focusing on TK, the teachers' CK in geography will influence the choice of pedagogy and technology as they consider what the key concepts are and how students can think geographically. This shift to focusing on how technology can contribute to the development of geographical knowledge and thinking has also been argued by Fargher (2017), Walshe (2018) and Healy and Walshe (2020).

A shift in focus to the intersection rather than only on the TK portion of the TPACK also alleviates some of the anxiety among geography teachers who have to design technology-based lessons. Chang and Wu (2018) introduced the concept of a prosumer of technology for geography teaching and argued that teachers do not need to be constantly updated, nor do they need high-level and sophisticated knowledge about technological affordances. Yet teachers should also not just accept and use standard forms of technology without considering the value in adapting their technology use for their unique contexts. Teachers will need some working knowledge of the functionality offered by technology in order to tailor their use of these technological tools for their own teaching contexts. Teachers can also focus on ensuring that the resources students use are meaningful in learning to know, learning to do, learning to be and learning to live together.

Fieldwork and geographical learning

The educational importance of geography fieldwork has been well established (Butt, 2020), and fieldwork is perceived as integral to geography education by geography academics and teachers alike (Foskett, 1999; Chang et al., 2018). Rickinson et al. (2004), based on a review of 150 pieces of research into outdoor learning, concluded that 'Substantial evidence exists to indicate that fieldwork properly conceived, adequately planned, well taught and effectively followed up, offers learners opportunities to develop their knowledge and skills' (p.5). Effective fieldwork has the potential to pique students' interest on geographical issues, encourage students to identify problems and pose questions, and form perceptions of changing landscapes. Moreover, fieldwork helps students to cultivate a synoptic understanding of geographical issues (Gerber and Goh, 2000; Chang et al., 2012).

Literature on fieldwork in geographical education has found that it helps students in integrating the theoretical and practical concepts learnt in the classroom and encourages holistic geographical understanding of issues (Kent et al., 1997). Fieldwork allows students the chance to make sense of their social, cultural and environmental landscapes (Gerber and Goh, 2000). Oost et al. (2011) recommend that fieldwork needs to be planned with consideration for both cognitive and affective learning, and the interaction between these. In order to capitalise upon the opportunities of fieldwork, we need to take the students beyond learning to know to learning to do, be and live together – to being a truly active, participating global citizen.

Unfortunately, some teachers conduct fieldwork as fieldtrips, which are in essence just excursions or tours where students are passive recipients of information (Chang and Ooi, 2008). Invariably, students describe these kinds of fieldtrips as boring and they have missed an opportunity to be deeply engaged in geographical fieldwork (Brown, 1969). I argue that fieldtrips that are well designed can provide students with 'experiences, knowledge, understanding as well as skills that are important to an understanding of the world around them' (Chang and Ooi, 2008, p.297). This focus on geographical experience, knowledge and understanding places an emphasis on the teachers' PCK when designing such learning activities.

Geography fieldtrips should provide a meaningful learning experience that are structured around an inquiry process of observation, discussion on the information collected and drawing conclusions based on that information. These processes should also be designed to allow a student to 'actively participate in discovering relevant places and making sense of them in the context of a particular topic' (Gaillard and McSherry, 2014, p.177). This resonates with the four pillars of the Delors (1996) report. The design of good geographical field inquiry is exemplified by the geographical investigation (GI) approach employed within the Singapore geography curriculum. In Singapore, a revised geography syllabus was implemented for the GCE 'O' levels in 2013 for Secondary 3 students (aged 14–15 years) (Chang and Seow, 2010), for Lower Secondary (students aged 12–14) in 2014 and GCE 'A' Levels in 2016 (Lim, 2018). A key feature of the revised syllabi was the use of the GI approach in the teaching and learning of geography to engage independent learning, critical thinking, reflective thinking and inquiry. In particular, the GI proposed was contingent on both fieldwork and classroom learning. The GI approach draws on the cycle of geographical enquiry advanced by Roberts (2003).

Robert's (2003) framework for learning through enquiry involves the four stages of:

1   Creating a need to know;
2   Using data;
3   Making sense; and
4   Reflecting on the learning.

(p.44)

Following these four stages, there is then opportunity for 'applying what has been learnt to next enquiry' (Roberts, 2003, p.44). Students investigate a geographical issue relating to specific topics in the syllabus. Each geographical issue has an accompanying GI question and students are then asked to collect, select and present relevant information from their field investigation. Based on this, the students form their own interpretations of the issue and provide an answer to the GI question (Lim, 2018, p.17). Lim (2018) examined the role of GI in developing students' cognitive thinking in geography for four students. All four students showed development of higher-order cognitive skills (e.g. analysis, evaluation and creation).

Seow et al. (2019) further argue that the Singapore GI can be considered a 'signature pedagogy' (Shulman, 2005) as a form of teaching that is commonly found within geography education in Singapore. Seow et al. (2019) suggest that GI can 'socialize students into the distinctive practices, concepts, and values of the discipline of geography' (p.227). Although the results suggest a need to develop 'teachers' place-based knowledge and relevant scientific content knowledge to support the design of field-based learning experiences for students' (Seow et al., 2019, p.227), it affirms the notion of the importance of PCK as argued earlier in this chapter.

Although there are clear benefits of using fieldwork in geographical education, there exist obstacles, at least in the perception of teachers, around factors such as bureaucracy, logistics, resources and time. For some teachers, it is a challenge to bring a large group of students to the field. Some teachers are also concerned with the costs involved in the fieldwork. As technologies and mobile technologies continue to advance, perhaps we can harness the affordances of technology to overcome these constraints.

Fieldwork with mobile technologies

A study by Favier and Van Der Schee (2009) identified successful examples of student research projects that combined quantitative 'data collection in the field with data visualisation, manipulation and analysis in GIS' (p.261). This certainly takes the students' learning beyond the knowing domain to the doing domain. In another example, 'playing Pokémon GO could develop students' spatial orientation ability in the same way as formal teaching courses specifically designed for the development of spatial orientation' (Carbonell Carrera et al., 2018, p.253). There have been many other studies on using mobile devices in fieldwork (Goh et al., 2007; Chatterjea et al., 2008; Nguyen et al., 2008; Razikin et al., 2009; Chang et al., 2012). Chatterjea (2018) describes the use of her own developed mobile application, the NIE mGEO for geographical fieldwork (in studying a forest reserve) in four stages:

1. Pre-fieldwork preparation.
2. Field observations and measurements, field assessment, and field decisions.
3. Post-field phase: data analysis, and research.
4. Knowledge-building phase: data assessment, data representation, and synthesis.

These stages have some resonance with Robert's (2003) stages of enquiry. There is still an imperative to create a need to know, collect and use the data (learning to do), and to make sense of the data and then reflect on the learning (learning to be). The key difference here is that the GI cycle was designed for a K-12 context while the stages described by Chatterjea (2018) are for university undergraduate and graduate students. Jarvis et al. (2016) emphasise that 'active learning that goes beyond data collection to encourage observation and thinking in the field is important' (p. 61), and this appears to be a dimension that needs to be capitalised upon when using mobile devices in the field.

Perhaps one could consider more explicitly the cross-fertilisation of ideas between academic and school geography. After all, the subject knowledge of geography teachers as applied in the school geography context is derived from their own geography learning as undergraduates. This is in alignment with the notion of TPACK where the subject knowledge that is learnt is not independent of the way in which the knowledge is acquired. How teachers themselves learn about geography and think about geography has an important bearing on the way students will be taught geography in schools. There is perhaps also the possibility that insights of technology and fieldwork use in school geography can inspire methods of teaching and learning geography in universities. Internationally, there has been increasing use of technology and field-based learning within university contexts (Welsh et al., 2013).

This chapter has argued that mobile technology use can be guided by the TPACK framework (Mishra and Koehler, 2006) and fieldwork enquiry can be guided by the enquiry approach (Roberts, 2003). This is not dissimilar to how TPACK and geographical enquiry approaches have been used in relation to GIS (e.g. Walshe, 2017; 2018; Healy and Walshe, 2020; Healy, 2021). The Delors (1996) report and the four pillars of learning to know, to do, to be and to live together also provide a framework for teachers to reflect on how meaningful geographical learning can be realised.

Final reflection

I am reminded of an old adage of six blind men who were led to an elephant. Each man was located at a different position relative to the elephant. The man touching the ears thought he was touching a huge leaf. The man touching the trunk thought he was holding a snake. The other four men all had different experiences, but they were all indeed touching the same animal. Learning geography need not only happen through one modality. Like the analogy of the blind men and the elephant, when we focus too much on what one modality can afford the learning experience, we may not consider how we can use an additional modality to support this same activity. This analogy reminds us that we need to consider how both mobile technologies and fieldwork, alongside other pedagogies, collectively support the teaching and learning of geography. At the same time, we cannot neglect the socio-emotional learning needs of our students. As we teach them to think critically about the data that they use, they should also consider their own cyber-wellness, as well as being considerate of the cyber-wellness of those around them (Mıhcı Türker and Kılıç Çakmak, 2019).

It is important to consider at this juncture the potential for both future research and practice in the use of technology for geography fieldwork. To only have 23 articles across two key geography education research journals over the last three decades is less than desirable. More empirical research needs to be done to inform new and innovative ways of using technology for fieldwork, especially when the rate at which technological advancement takes place is only going to increase (Kurzweil, 2006). With the growing potential of big data and artificial intelligence innovations, the use of technology in fieldwork will likely take new forms that we cannot even imagine.

References

Bednarz, R.S. and Bednarz, S.W., 2004. Geography education: The glass is half full and it's getting fuller. *The Professional Geographer*, 56 (1), pp.22–27.

Béneker, T. and van der Schee, J., 2015. Future geographies and geography education. *International Research in Geographical and Environmental Education*, 24 (3), pp.287–293.

Brown, E.H., 1969. The teaching of fieldwork and the integration of physical geography. In R.U. Cooke and J.H. Johnson (eds), *Trends in Geography: An Introductory Survey*. London: Heinemann, pp.70–78.

Butt, G., 2020. *Research in Geography and Geography Education: The Roles of Theory and Thought*. Cham: Springer.

Carbonell Carrera, C., Saorín, J.L. and Hess Medler, S., 2018. Pokémon GO and improvement in spatial orientation skills. *Journal of Geography*, 117 (6), pp.245–253.

Chang, C.H. and Ooi, G.L., 2008. Role of fieldwork in humanities and social studies education. In O.S. Tan, D.M. McInerney, A.D. Liem and A.-G. Tan (eds), *What the West Can Learn from the East: Asian Perspectives on the Psychology of Learning and Motivation*. Charlotte, NC: Information Age Publishing, pp.295–312.

Chang, C.H. and Seow, T., 2010. Field inquiry for Singapore geography teachers. In *SEAGA International Conference 2010. Hanoi, Vietnam, 23–26 Nov 2010*. Singapore: Southeast Asian Geography Association.

Chang, C.H. and Wu, B.S., 2018. Teaching geography with technology: A critical commentary. In C.H. Chang, B.S. Wu, T. Seow and K. Irwine (eds), *Learning Geography Beyond the Traditional Classroom*. Singapore: Springer, pp.35–47.

Chang, C.H., Chatterjea, K., Goh, D.H.L., Theng, Y.L., Lim, E.P., Sun, A., Razikin, K., Kim, T.N.Q. and Nguyen, Q.M., 2012. Lessons from learner experiences in a field-based inquiry in geography using mobile devices. *International Research in Geographical and Environmental Education*, 21 (1), pp.41–58.

Chang, C.H., Irvine, K., Wu, B.S. and Seow, T., 2018. Reflecting on field-based and technology-enabled learning in geography. In C.H. Chang, B.S. Wu, T. Seow and K. Irwine (eds), *Learning Geography Beyond the Traditional Classroom*. Singapore: Springer, pp.201–212.

Chatterjea, K., 2018. Authentic learning: Making sense of the real environment using mobile technology tool. In C.H. Chang, B.S. Wu, T. Seow and K. Irwine (eds), *Learning Geography Beyond the Traditional Classroom*. Singapore: Springer, pp.111–131.

Chatterjea, K., Chang, C.H., Lim, E.P., Zhang, J., Theng, Y.L. and Go, D.H.L., 2008. Supporting holistic understanding of geographical problems: Fieldwork and G-Portal. *International Research in Geographical and Environmental Education*, 17 (4), pp.330–343.

Delors, J., 1996. Learning: The treasure within (Report to UNESCO on the International Commission on Education for the Twenty-First Century, Paris UNESCO 1996). *Internationales Jahrbuch der Erwachsenenbildung*, 24 (1), pp.253–258.

Fargher, M., 2017. GIS and the power of geographical thinking. In C. Brooks, G. Butt and M. Fargher (eds), *The Power of Geographical Thinking*. Switzerland: Springer, pp.151–164.

Favier, T. and Van Der Schee, J., 2009. Learning geography by combining fieldwork with GIS. *International Research in Geographical and Environmental Education*, 18 (4), pp.261–274.

Foskett, N., 1999. Forum: Fieldwork in the geography curriculum – international perspectives and research issues. *International Research in Geographical and Environmental Education*, 8 (2), pp.159–163.

Gaillard, J.C. and McSherry, A., 2014. Revisiting geography field trips: A treasure hunt experience. *Journal of Geography*, 113 (4), pp.171–178.

Gerber, R. and Goh, K.C. (eds), 2000. *Fieldwork in Geography: Reflections, Perspectives and Actions*. Dordrecht: Kluwer Academic.

Goh, D.H.L., 2007. Learning geography with the G-portal digital library. In A. Tatnall (ed.), *Encyclopedia of Portal Technologies and Applications*. London: IGI Global, pp.547–553.

Goodchild, M.F., 2007. Citizens as sensors: The world of volunteered geography. *GeoJournal*, 69 (4), pp.211–221.

Hammond, L., 2021. Children, childhood, children's geographies: Evolving through technology. In N. Walshe and G. Healy (eds), *Geography Education in the Digital World*. London: Routledge, pp.38–49.

Healy, G., 2021. Insights from professional discourse on GIS: A case for recognising geography teachers' repertoire of experience. In N. Walshe and G. Healy (eds), *Geography Education in the Digital World*. London: Routledge, pp.89–101.

Healy, G. and Walshe, N., 2020. Real-world geographers and geography students using GIS: Relevance, everyday applications and the development of geographical knowledge. *International Research in Geographical and Environmental Education*, 29 (2), pp.178–196.

Hsu, T.Y. and Chen, C.M., 2010. A mobile learning module for high school fieldwork. *Journal of Geography*, 109 (4), pp.141–149.

International Geographic Union, Commission on Geographical Education (IGU-CGE), 1992. *International Charter on Geographical Education*. International Geographical Union, Commission on Geographical Education. Available at: www.igu-cge.org/wp-content/uploads/2018/02/1.-English.pdf [Accessed 31 May 2019].

International Geographic Union, Commission on Geographical Education(IGU-CGE), 2016. *International Charter on Geographical Education*. International Geographical Union, Commission on Geographical Education. Available at: www.igu-cge.org/wp-content/uploads/2019/03/IGU_2016_eng_ver25Feb2019.pdf [Accessed 31 May 2019].

Jarvis, C., Tate, N., Dickie, J. and Brown, G., 2016. Mobile learning in a human geography field course. *Journal of Geography*, 115 (2), pp.61–71.

Kent, M., Gilbertson, D.D. and Hunt, C.O., 1997. Fieldwork in geography teaching: A critical review of the literature and approaches. *Journal of Geography in Higher Education*, 21 (3), pp.313–332.

Kitchen, B., 2021. Using mobile virtual reality to enhance fieldwork experiences in school geography. In N. Walshe and G. Healy (eds), *Geography Education in the Digital World*. London: Routledge, pp.131–141.

Kurzweil, R., 2006. *The Singularity is Near*. Richmond: Duckworth.

Lambert, D., 2016. Geography. In D. Wyse, L. Hayward and J. Pandya (eds), *The Sage Handbook of Curriculum, Pedagogy and Assessment*. London: Sage, pp.391–407.

Laws, K., 1989. Learning geography through fieldwork. *The Geography Teachers' Guide to the Classroom*, 2, pp.104–117.

Lim, E.Q., 2018. The role of geographical investigations in developing students' cognitive thinking. *HSSE Online – Research and Practice in Humanities and Social Studies Education*, 7 (1), pp.14–27.

Medzini, A., Meishar-Tal, H. and Sneh, Y., 2015. Use of mobile technologies as support tools for geography field trips. *International Research in Geographical and Environmental Education*, 24 (1), pp.13–23.

Mishra, P. and Koehler, M.J., 2006. Technological pedagogical content knowledge: A framework for teacher knowledge. *Teachers College Record*, 108 (6), pp.1017–1054.

Morgan, J., 2019. *Culture and the Political Economy of Schooling: What's Left for Education?* Abingdon: Routledge.

Nguyen, Q.M., Kim, T.N.Q., Goh, D.H.L., Theng, Y.L., Lim, E.P., Sun, A., Chang, C.H. and Chatterjea, K., 2008. TagNSearch: Searching and navigating geo-referenced collections of photographs. In *ECDL, Research and Advanced Technology for Digital Libraries: 12th European Conference. Aarhus, Denmark, 14–19 September 2008*. Berlin, Heidelberg: Springer, pp.62–73.

Oost, K., De Vries, B. and Van der Schee, J.A., 2011. Enquiry-driven fieldwork as a rich and powerful teaching strategy – school practices in secondary geography education in the Netherlands. *International Research in Geographical and Environmental Education*, 20 (4), pp.309–325.

O'Reilly, T., 2007. What is Web 2.0: Design patterns and business models for the next generation of software. *Communications and Strategies*, 1, pp.17–37.

Pike, S., 2021. GIS for young people's participatory geography. In N. Walshe and G. Healy (eds), *Geography Education in the Digital World*. London: Routledge, pp.117–128.

Priestnall, G., 2021. Augmented reality: Opportunities and challenges. In N. Walshe and G. Healy (eds), *Geography Education in the Digital World*. London: Routledge, pp.155–167.

Razikin, K., Goh, D.H.L., Theng, Y.L., Nguyen, Q.M., Kim, T.N.Q., Lim, E.P., Chang, C.H., Chatterjea, K. and Sun, A., 2009. Sharing mobile multimedia annotations to support inquiry-based learning using MobiTOP. In *AMT, International Conference on Active Media Technology. Beijing., China. 22–24 October 2009*. Berlin, Heidelberg: Springer, pp.171–182.

Rickinson, M., Dillon, J., Teamey, K., Morris, M., Choi, M.Y., Sanders, D. and Benefield, P., 2004. *A Review on Outdoor Learning*. Shrewsbury: Field Studies Council.

Roberts, M., 2003. *Learning through Enquiry: Making Sense of Geography in the Key Stage 3 Classroom*. Sheffield: Geographical Association.

Roberts, M., 2021. Geographical sources in the digital world: Disinformation, representation and reliability. In N. Walshe and G. Healy (eds), *Geography Education in the Digital World*. London: Routledge, pp.53–64.

Schaal, S., 2021. Location-based games for geography and environmental education. In N. Walshe and G. Healy (eds), *Geography Education in the Digital World*. London: Routledge, pp.168–177.

Seow, T., Chang, J. and Irvine, K.N., 2019. Field-based inquiry as a signature pedagogy for geography in Singapore. *Journal of Geography*, 118 (6), pp.227–237.

Shulman, L.S., 1986. Those who understand: Knowledge growth in teaching. *Educational Researcher*, 15 (2), pp.4–14.

Shulman, L.S., 2005. Pedagogies of uncertainty. *Liberal Education*, 91 (2), pp.18–25. Türker, P.M. and Çakmak, E.K., 2019. An investigation of cyber wellness awareness: Turkey secondary school students, teachers, and parents. *Computers in the Schools: Interdisciplinary Journal of Practice, Theory, and Applied Research*, 36 (4), pp.293–318.

Mıhcı Türker, P. and Kılıç Çakmak, E., 2019. An Investigation of Cyber Wellness Awareness: Turkey Secondary School Students, Teachers, and Parents. *Computers in the Schools*, 36 (4), pp.293–318.

Walshe, N., 2017. Developing trainee teacher practice with geographical information systems (GIS). *Journal of Geography in Higher Education*, 41 (4), pp.608–628.

Walshe, N., 2018. Spotlight on ... Geographical information systems for school geography. *Geography*, 103 (1), pp. 46–49.

Welsh, K.E., Mauchline, A.L., Park, J.R., Whalley, W.B. and France, D., 2013. Enhancing fieldwork learning with technology: Practitioner's perspectives. *Journal of Geography in Higher Education*, 37 (3), pp.399–415.

Wu, B.S., 2013. Developing an evaluation framework of spatial understanding through GIS analysis of volunteered geographic information (VGI). *Review of International Geographical Education Online*, 3 (2), pp.152–162.

Yuen, S.C.Y., Yaoyuneyong, G. and Johnson, E., 2011. Augmented reality: An overview and five directions for AR in education. *Journal of Educational Technology Development and Exchange (JETDE)*, 4 (1), pp.119–140.

Annex A: Articles reviewed for the analysis in Table 13.1

Bartlett, K., 1999. Field sketching: An appropriate skill for upper primary children? *International Research in Geographical and Environmental Education*, 8 (2), pp.199–207.

Barton, K., 2017. Exploring the benefits of field trips in a food geography course. *Journal of Geography*, 116 (6), pp.237–249.

Carbonell Carrera, C., Saorín, J.L. and Hess Medler, S., 2018. Pokémon GO and improvement in spatial orientation skills. *Journal of Geography*, 117 (6), pp.245–253.

Chang, C.H., Chatterjea, K., Goh, D.H.L., Theng, Y.L., Lim, E.P., Sun, A., Razkin, K., Kim, T.H.Q. and Nguyen, Q.M., 2012. Lessons from learner experiences in a field-based inquiry in geography using mobile devices. *International Research in Geographical and Environmental Education*, 21 (1), pp.41–58.

Favier, T. and van Der Schee, J., 2009. Learning geography by combining fieldwork with GIS. *International Research in Geographical and Environmental Education*, 18 (4), pp.261–274.

Gaillard, J.C. and McSherry, A., 2014. Revisiting geography field trips: A treasure hunt experience. *Journal of Geography*, 113 (4), pp.171–178.

Hsu, T.Y. and Chen, C.M., 2010. A mobile learning module for high school fieldwork. *Journal of Geography*, 109 (4), pp.141–149.

Jarvis, C., Tate, N., Dickie, J. and Brown, G., 2016. Mobile learning in a human geography field course. *Journal of Geography*, 115 (2), pp.61–71.

Kastens, K.A. and Liben, L.S., 2010. Children's strategies and difficulties while using a map to record locations in an outdoor environment. *International Research in Geographical and Environmental Education*, 19 (4), pp.315–340.

Krakowka, A.R. 2012. Field trips as valuable learning experiences in geography courses. *Journal of Geography*, 111 (6), pp.236–244.

Leydon, J. and Turner, S., 2013. The challenges and rewards of introducing field trips into a large introductory geography class. *Journal of Geography*, 112 (6), pp.248–261.

Medzini, A., Meishar-Tal, H. and Sneh, Y., 2015. Use of mobile technologies as support tools for geography field trips. *International Research in Geographical and Environmental Education*, 24 (1), pp.13–23.

Papadimitriou, F., 2010. Introduction to the complex Geospatial Web in geographical education. *International Research in Geographical and Environmental Education*, 19 (1), pp.53–56.

Remmen, K.B. and Frøyland, M., 2014. Implementation of guidelines for effective fieldwork designs: Exploring learning activities, learning processes, and student engagement in the classroom and the field. *International Research in Geographical and Environmental Education*, 23 (2), pp.103–125.

Remmen, K.B. and Frøyland, M., 2015. What happens in classrooms after earth science fieldwork? Supporting student learning processes during follow-up activities. *International Research in Geographical and Environmental Education*, 24 (1), pp.24–42.

Robertson, M., 2009. Young 'netizens' creating public citizenship in cyberspace. *International Research in Geographical and Environmental Education*, 18 (4), pp.287–293.

Rundquist, B.C. and Vandeberg, G.S., 2013. Fully engaging students in the remote sensing process through field experience. *Journal of Geography*, 112 (6), pp.262–270.

Rydant, A.L., Shiplee, B.A., Smith, J.P. and Middlekauff, B.D., 2010. Applying sequential fieldwork skills across two international field courses. *Journal of Geography*, 109 (6), pp.221–232.

Sheppard, P.R., Donaldson, B.A. and Huckleberry, G., 2010. Quantitative assessment of a field-based course on integrative geology, ecology and cultural history. *International Research in Geographical and Environmental Education*, 19 (4), pp.295–313. Sinha, G., Smucker, T.A., Lovell, E.J., Velempini, K., Miller, S.A., Weiner, D. and Wangui, E.E., 2017. The pedagogical benefits of participatory GIS for geographic education. *Journal of Geography*, 116 (4), pp.165–179.

Skop, E. 2009. Creating field trip-based learning communities. *Journal of Geography*, 107 (6), pp.230–235.

Tal, T., 2010. Pre-service teachers' reflections on awareness and knowledge following active learning in environmental education. *International Research in Geographical and Environmental Education*, 19 (4), pp.263–276.

Walsh, M.K., 2014. Teaching geographic field methods using paleoecology. *Journal of Geography*, 113 (3), pp.97–106.

Chapter 14

Augmented reality
Opportunities and challenges

Gary Priestnall

Introduction

The term augmented reality (AR) was originally used to refer to overlaying computer-generated graphics onto objects in the real-world scene, but today it can take on a broader meaning. It could be considered to be any technology which enhances our understanding of the physical environment using *in situ* digital content and, as such, is inherently geographical. This opens up exciting possibilities for designing learning experiences which combine physical movement and exploration with the richness of digital multimedia. As an emerging technology, AR has ridden waves of hype for many years but has yet to realise its potential. Adopting a broad definition of AR, this chapter presents a structured summary of the technological approaches to augmenting reality and considers some of the opportunities for its use in a geographical context. It also reflects on a number of research challenges that remain in developing and deploying effective AR techniques, particularly in educational contexts, drawing on experiences of augmenting the landscape on fieldtrips and in the use of projection augmentation on landscape models in the classroom. The next section offers a new categorisation of AR, based on a very broad definition, in order to raise awareness of the breadth of options that might be considered by teachers of geography.

Techniques for augmenting reality

Whilst virtual reality (VR) attempts to immerse the user in an entirely virtual world, AR enhances the user's experience in the real world through digital media. Between the entirely real and the entirely virtual are technologies that offer a range of 'mixed realities', termed the 'reality-virtuality continuum' by Milgram and Kishino (1994). Early definitions of AR emphasised a strong connection to the virtual world, suggesting that there should be accurate registration between real and virtual objects in three-dimensional space (Azuma, 1997) using an array of high-end sensors. As more techniques were developed for bringing virtual content into real-world situations, using visual markers to trigger content for example, AR became more of a general concept than a very specific technological configuration. With the arrival of mobile devices, the concept gained breadth of application and saw AR being used as a term to refer to location-triggered multimedia.

Despite the rapidly increasing number of AR case studies in technology-enhanced learning journals, as reported by Hwang and Wu (2014), the use of AR in the classroom is not widespread. There is, however, now a wide range of techniques that could be classed as AR in its broadest sense, varying in both technical detail and capability. Table 14.1 summarises the breadth of approaches taken for augmenting reality, using a structure which is followed

Table 14.1 Categories of augmented reality.

Category of AR	Description
Sensor-based	Sensors including inertia devices are used to attempt to align digital content with features in the real world to which the content relates
Immersive	Users experience content through some kind of head mounted display
Non-immersive	Users experience content via a mobile device
Marker-based	Content is triggered by some kind of visual pattern in the scene using the camera and image processing software on a mobile device
Code	Some kind of two-dimensional barcode, typically made up of black and white squares, as with a Quick Response (QR) code
Recognisable patterns	Patterns which are either more aesthetically pleasing or actually mean something to the viewer can also be used to create 'machine-readable' labels for triggering AR content
Location-based	Content is triggered by some kind of positioning sensor, termed 'locative media'
Place-specific	Content relates to the location where it is triggered
Remote	Content does not relate to the place it is triggered and is either more abstract or relates to a remote landscape, often at a different scale to the one used to trigger it
Spatially based	Spatial AR uses projection to augment elements of the real world scene
Objects	Objects in the environment are augmented with 'texture maps'
Landscape model	A physical scale model of a landscape is textured with images, maps and animations

in the remainder of this chapter to review the technologies, with a particular emphasis on potential applications in geography education.

Sensor-based

Immersive

The first AR systems used head mounted displays (HMD) to overlay computer-generated graphics over objects in the scene. As well as providing a visually immersive experience, the real-time interaction with the scene was very natural due to movement being controlled by an inertia device on the user's head. The technology developed from lab-based devices to outdoor systems with the addition of high accuracy GPS positioning, an example being the Tinmith outdoor AR system (Piekarski and Thomas, 2003) which merged computer-generated graphics with a video stream of the user's view of the scene. The ability to superimpose detailed 3D models accurately onto the scene has clear potential for reconstructing past environments, as illustrated by Vlahakis et al. (2002) in the context of archaeological sites. In this case reconstructions of ancient temples were merged with live video streams of the present day scenes displayed through the user's HMD. This was a technically complex prototype but had the advantage of aligning reconstructions dynamically as the user moved their head. An interesting development more recently has been the HoloLens from Microsoft which uses sensors to detect objects in the scene to enable 3D models to be placed in the environment. From a geographical perspective the potential to visualise digital landscape models as if they were tabletop displays offers exciting possibilities, for example in the context of visualising urban dynamics (Chen et al., 2017).

Non-immersive

The emergence of powerful tablet computers and smartphones have allowed some sensor-based approaches to be developed which do not rely on high-end technologies, but instead use the device screen rather than the HMD, so lack visual immersion. Developments in these mobile AR systems, which exploit accelerometer, magnetometer and gyroscope sensors, were reviewed by Jamali et al. (2014). For example, one technique uses the camera in a phone, in combination with its motion sensors, to detect planes in the local scene onto which computer-generated models can be dropped; examples of this include Apple's ARKit[1] and Google's ARCore.[2]

Rather than attempting to place the geometry of virtual 3D models accurately into real-world scenes, some approaches to mobile AR use symbols floating over the scene via the camera view on the mobile device. These could be termed 'AR browsers' in that they allow local information to be searched and presented as thematic layers made up of points of interest (PoIs). Examples here included Wikitude (www.wikitude.com) and Layar (www.layar.com). In a fieldtrip context these could display information about the landscape around the user, for example showing houses for sale nearby by dropping labels over the camera view of the mobile device as it was pointed in different directions. Mobile AR offers some potential in the context of geographical fieldwork to enrich the user's experience in a location, although we must be aware of the influence of such approaches imposing certain narratives on a place, as addressed by Liao and Humphreys (2015). AR authoring tools open up possibilities for creating interesting field exercises which engage students with the process of attaching meaning to landscape, in addition to supplying what might be deemed 'official' descriptions of PoIs within those landscapes. For example, if students are able to create descriptions of what a particular part of a town means to them, and make these available to other students to experience in the field as part of a group exercise, then it could promote discussion about various and even contested notions of place.

Marker-based

Codes

Using machine vision to recognise visual codes in the camera view of a mobile device can offer a simple way to associate virtual content with that location, be it on the page of a book or in an outdoor scene. Often such codes are black and white two-dimensional barcodes like quick response (QR) codes, and so provide an obvious cue to users that there is information to be explored at that point. Physical markers like this can be placed at locations in outdoor environments to control the triggering of rich media related to that spot in the landscape, as with an ecological game developed by Hwang et al. (2016). The media does not always have to relate to the location of the code, for example in the marine ecology game described by Lu and Liu (2015) media is triggered by pointing the camera at codes printed on the shirts of instructors. On other occasions the media triggered by the code needs to be aligned with the code itself, as seen in Carbonell et al. (2018), where 3D block diagrams of landscapes are revealed alongside contour maps of the same landscape.

Recognisable patterns

Whilst codes provide obvious visual cues for exploring augmented content they do not convey any information directly to the human eye, nor can they be seen as representing a particular theme of interest or anything that would be considered useful outside of the AR experience. Another branch of AR, therefore, concerns itself with using more recognisable patterns within the visual scene to trigger content. An example of this technique would be Artcodes (Benford et al., 2017) where visually interesting patterns can be designed to contain machine readable codes. These codes have certain visual structures built in which encode topological relationships rather than the exact geometries of QR codes. Another approach is to enable parts of images or objects to become machine-readable patterns which can trigger media related to those features. Such patterns have been termed 'auras', and could be parts of paintings in a gallery (Clini et al., 2014), exhibits in an informal learning setting (Sommerauer and Müller, 2014) or sculptures in an outdoor setting (Bower et al., 2014). The creation of cheap and easy authoring environments to allow drawings to be associated with multimedia content clearly has potential for creating engaging learning experiences both in the classroom and in the field. For example, if a teacher created a series of images to act as the codes, printed them out and placed them around a space, they could be used to trigger relevant media files as students moved around to discover them. They could be placed next to physical objects to which the media might relate, or simply distributed in a pattern that would encourage exploration.

Location-based

Place-specific

A class of mobile technologies that use location as the primary trigger to reveal media, often termed 'locative media', has been used to generate experiences referred to as 'augmented reality' in the broader sense. Location sensing can be achieved in small spaces using RFID (radio frequency identification) tags and a wireless communications network (Hwang et al., 2011) or in the wider landscape using GPS (global positioning system) to create 'mediascapes' (Stenton et al., 2007). The mediascape approach offers interesting opportunities to augment the field experience with useful media, even if such media does not relate directly to elements of the visual scene, unlike conventional AR. For example the augmented reality learning approach described by Efstathiou et al. (2018) used location-based multimedia to promote engagement with, and learning about, an archaeological site. In this case the media at each 'hotspot' was organised around a consistent structure, with themed sub-sections offering a mix of video, text and imagery, describing the activities that would have occurred at that location. Locative media could also be used to encourage students to think about the relationship between space and place and the alternative associations people might have with a particular spatial location (Dourish, 2006).

There is scope to encourage students to think about the relationship between habitats, organisms and physical processes in the field using locative media, for example in the Ambient Wood project (Rogers et al., 2004). Here a framework for digital augmentation was presented which included contextually relevant information being made available to students, either on demand or in response to their movement through the wood. The exercise also featured a remote facilitator to assist with the recognition of species and to encourage in-field reflection. The field experience can also be augmented by instructions for taking measurements as with the EcoMOBILE project (Kamarainen et al., 2013) or for simulating measurements virtually as with Squire and Klopfer (2007) in relation to contaminated groundwater.

Location triggers can be used to project media relating more directly to specific views onto the landscape (Priestnall et al., 2019), in so doing offering a form of AR where virtual content and real scene are more loosely coupled than in sensor-based AR. Whilst this simple technique offers many possibilities for augmenting scenes with reconstructions or annotated views of the same scene, it is limited by having to author the media in advance. Gill and Lange (2015) illustrate a system whereby views are projected onto a scene that can be dynamically generated and viewed from any desired viewing position whilst in the field in the context of urban planning.

A major challenge of location-based augmentation is in directing the gaze of the user towards the part of the landscape scene to which the media relates. For example, in the context of geographical fieldwork where the media supplied might relate to parts of the landscape scene around the user, such as buildings, mountains or parts of a river, rather than to their immediate location. Bartie and Mackaness (2006) modelled the visibility of landmark buildings, along with the direction of travel from recent GPS locations, to deliver audio to the user which directed their gaze towards the building of interest before describing it. Some research developments have attempted to engage more directly with the landscape scene, utilising orientation sensors in the device, as well as GPS, to allow the user to point the phone at parts of the landscape scene to receive information; for example the *geowand* approaches of Carswell et al. (2010) and Meek et al. (2013). In such cases the direction in which the device is being pointed is calculated using the sensors within the device, and this is then used in conjunction with the current location of the device to intersect with objects in a geographical database.

Remote

A less common but nevertheless interesting use of locative media has been to use movement through space to trigger information about a different landscape, often one which is remote to the user and frequently much larger than the study area used as the backdrop to the exercise. For example, Facer et al. (2004) used a playing field measuring 100m x 50m as the basis for a locative media exercise to allow students to explore the sights and sounds of the African Savannah landscape. In a similar way, Bursztyn et al. (2017) used part of a college campus to explore the geological history of the Grand Canyon by triggering media through physical movement around that space. Such approaches could be used to promote awareness of the geographical similarities and differences between a student's local area and a region in a remote part of the world, for example the 'place knowledge' component of the Geography National Curriculum of England.

Spatially based

Objects

A distinctly different form of AR is where elements of the real-world scene are augmented through direct projection, rather than viewing augmentation through a screen. This technique has been termed 'spatial augmented reality' (Bimber and Raskar, 2005) and uses projection to augment physical objects with texture map images, lighting effects and animations.

Landscape models

A form of spatial AR that has relevance to geographical education uses landscape models as the focus for projection. One technique is to use malleable models that can be augmented with projected contours, water flow and other landscape variables. Early examples are the Illuminating Clay project and its follow-on developments (Piper et al., 2002; Mitsova et al., 2006) where small clay landscape models were manipulated by hand resulting in the new surface being detected and modelled in real-time. Representations of the newly re-formed surface, including contour maps and water flow simulations, were projected down onto the model. Such techniques are highly engaging and allow simple form-process relationships to be modelled. More recent developments of this technique have employed sand rather than clay, termed 'AR sandboxes', which have gained popularity as engaging public displays and have been demonstrated as useful in educational settings (Woods et al., 2016). Such systems can be built following instructions online[3] using a data projector and computer, though due to the heavy computational requirements it will be necessary to have a powerful graphics card.

Whilst AR sandboxes offer engaging and highly interactive experiences they do not represent real places. The Projection Augmented Relief Model (PARM) technique (Priestnall et al., 2012) uses solid relief models augmented with projection to convey detailed geographic information, typically based on data exported from geographical information systems (GIS). Using this technique it is possible to allow students to interactively display various layers of spatial information using the natural frame of reference offered by the physical relief model.

The next two sections offer case studies to reflect on two very different forms of augmentation. The first case study focuses on a place-specific form of location-based augmentation using positional triggers to reveal context-relevant information in a field exercise. The second case study uses a projection augmented landscape model in the classroom to provide an engaging tangible display for studying geographical patterns in space and time over a large area.

Case study 1: Experiences from using field-based augmentation

In order to reflect on the process of augmenting the field experience using digital media, I will summarise some experiences from a number of past field exercises. In terms of the classification presented above, this represents an example of place-specific location-based augmentation. Early experiments used printed acetates (Priestnall, 2009) in order to compare annotated wireframes from digital landscapes with their real-world counterparts in upland North Wales. As the digital representation was not being used directly in conjunction with the real scene this would not be considered AR; however, it did engage students in the relationship between the digital and the real, and offered an opportunity in later fieldtrips to try out ideas for the design of augmented displays. It may also be worth considering for teachers who wish to experiment with a low-tech form of augmentation, for example by creating labelled overlays using photographs as a guide, and then printing just the overlay onto an acetate.

A range of techniques including acetate, bespoke and off-the-shelf locative media apps, and immersive in-field VR were employed across a number of fieldtrips to provide a focus for a university level geospatial technology module (Priestnall et al., 2019). Simplicity of design in terms of the user experience was important, but in the context of this technology-focused module there were opportunities to engage students with issues relating to GPS accuracy and, in the case of the locative media augmentations, the effectiveness of various shapes, sizes and positions of trigger zones. It became apparent through practice that it was the combination of trigger zone, GPS accuracy and design of the media that influenced success, in terms of whether the user found the media useful in augmenting the landscape scene. With orientation sensors not being employed in locative media approaches it was up to the user to make the association between media and landscape, and this could be disrupted if the media triggered too early and there were no explicit cues built into the media to aid orientation. An aspect of the design of the exercise which proved successful was letting groups of students author their own mediascapes and then go out in the field and test the mediascapes of another group.

A more general observation was that the deployment of technologies for augmenting the real world encouraged students to observe the landscape in some detail. It also gave an opportunity to explore a mix of media relating to both physical and cultural geographies, and in so doing promoted an awareness of different ways of seeing the landscape. This was also seen when using mediascapes with Year 6 students in Clifton village, Nottingham. The aim of the exercise was to engage students in the historical geography of the village through a treasure hunt style exploration. Drawing upon experiences with authoring and testing trigger zones and media on the university geospatial technologies field exercise, media used with the school employed quite explicit references to orientation, using instructions and images to gamify the process of finding the right location, as shown in Figure 14.1. By having to find out information about a number of sites across the village, and to identify specific objects and viewpoints, the exercise appeared to be successful in promoting spatial thinking about the development of the village. The use of maps on the mediascape app, and also printed maps, was seen to be important in helping to orientate the students to the places of interest, and the gamification of finding the matching views was popular, prompting interesting group discussions.

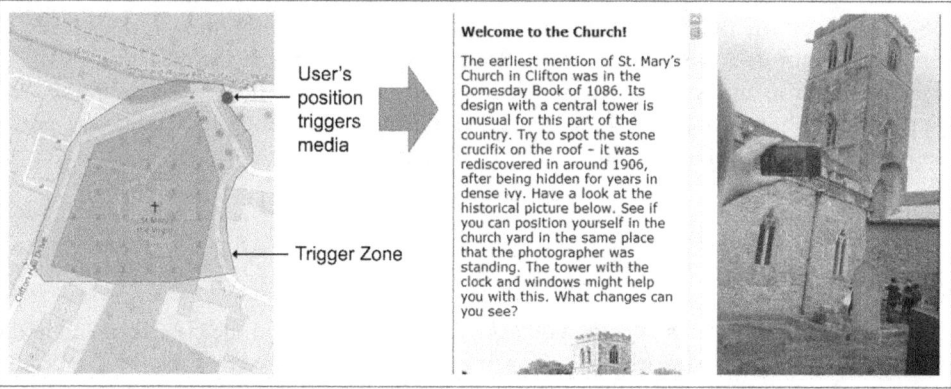

Figure 14.1 Augmented scenes using locative media.

Case study 2: Experiences from deploying projection augmented landscape models

Experiences with using digital augmentation in the classroom have been gained from deploying PARM displays in the classroom from Year 6 through to Year 13 in various contexts. The displays feature 3D printed models of landscapes, derived from airborne laser-scanning data, augmented through projected maps, images and animations. There are a number of affordances of PARM displays which suggest they could be beneficial in a teaching and learning context:

- The physicality of the model offers a natural frame of reference for identifying landscape features and their spatial arrangement and so may form an effective base for presenting spatial information as an alternative to a map.
- The projection-augmentation gives a holographic effect which offers a novel and engaging type of display as a focus for attention.
- The ability to touch and trace features on the model and to move around it changing the viewing perspective offer kinaesthetic qualities which appear to maintain interest.
- PARM has proved successful as a device for telling stories about landscapes in museum and visitor centre contexts, and so with easy authoring using Microsoft PowerPoint it allows students to create presentations from a series of digital data overlays.
- PARM has been seen to promote discussion in groups when in public contexts so could form a focus for group working around geographical problems.

Focusing on the use of a model of Keswick in Cumbria, UK, this offered a platform for Year 10 students to work in groups to develop stories related to the theme of 'changing places', using digital map overlays and annotations in PowerPoint to present the development of the town from a certain perspective. It proved useful in engaging students with the concept of GIS-like data layers, but it was also a hugely popular mode of presentation, with students demonstrating creativity with animated annotations and interesting transitions between layers.

For Year 12 groups the same model was used in the context of flood risk as shown in Figure 14.2. In this case the display featured a series of pre-generated sections including a series of historic map overlays to show the development of the town, maps of the flood defences, and several animated sequences modelling the extent of the flood hazard over a 48 hour period. As the flood extent maps were animated across the model, the passing of time was shown as a moving bar over a flood hydrograph on a monitor. The PARM was useful not only as an instructor-led display device to describe the flood problem in the area, but also as an interactive tool for group discussion to support the production of a poster summarising the nature of the flood hazard in Keswick.

Discussion

Opportunities

A general observation from exploring digital augmentation, both in the classroom and in the field, is that all technologies falling within the broad range described in Table 14.1 appear to facilitate multimedia learning in interesting ways. The notion of linking text, images and other media closely to support learning, as described by Moreno and Mayer (1999), could

Figure 14.2 Projection Augmented Relief Model (PARM) supporting in-class discussion of flood risk.

be seen to be a fundamental part of AR, which exploits the power of mixed media in different contexts. Typically the learner is more active in exploring media via AR, be that searching out codes in a classroom to reveal animated 3D content, or moving around an outdoor environment to trigger new ways of seeing a landscape.

The power of all forms of AR to label an apparent single reality with multiple, alternative and even contested geographies has applications across the breadth of geographical education. This could be revealing alternative physical properties of an outdoor landscape in situ, or playing out a series of geographical patterns over time projected onto a physical model of a landscape. This relates to the often complex interplay between the many physical, historical and cultural factors that shape a landscape. AR has the potential to allow users to play a more active role in exploring these alternative geographies whilst situated in the landscape itself. There is also, in most cases, scope to raise awareness of digital geographic information layers, and in so doing make connections to the capabilities of GIS.

All forms of AR appear to promote a kinaesthetic mode of learning. Indoor AR, whether a projection-enhanced model or codes placed on objects around a room, requires physical movement and interaction to reveal information designed to support learning. In outdoor settings there is even more potential to create learning experiences which promote physical exploration and to exploit digital content in the field, rather than back in the field centre. Digital content can play an important role both before and after a fieldtrip (Dykes et al., 1999; McMorrow, 2005), but AR presents various options for integrating digital media into the field experience. In particular it offers the potential to do so in a way that encourages physical exploration in order to discover content and also mechanisms to encourage users to relate that content to relevant parts of the landscape. An important virtue of AR is to reveal non-visible geographies superimposed in some way over the visible landscape, include past environmental conditions but also any variable for which there are spatial distributions, from pollution levels to perceptions of crime. By its very nature AR offers exciting possibilities to enhance fieldwork rather than offer replacements for it (McCauley, 2017), though we should continue to reflect critically on the effectiveness of the technologies to ensure they are not getting in the way of learning.

Challenges

AR has been in the public's consciousness for some time and has experienced a wave of hype which could potentially slow further adoption. According to the research and advisory company Gartner, AR first appeared on the rise in the 'emerging technologies hype cycle' in 2005 largely due to the innovation trigger of smartphones with on-board sensors enabling mobile AR. By 2011 AR was at the top of the 'peak of inflated expectations', but by 2018 it was at the bottom of the 'trough of disillusionment'. The launch of Pokémon Go in 2016 brought the concept of AR back into the public eye for a time, and whilst it was seen to promote outdoor exploration it also generated concern over the nature of the geographical movements it promoted (Colley et al., 2017). By 2019 AR had disappeared from the hype cycle, meaning it had matured and was no longer considered to be an emerging technology; this was perhaps helped by commercial platforms, such as Apple's ARKit and Google's ARCore. In terms of educational technology there is still scope for the development of authoring environments to make content creation for both students and teachers easier. Off-the-shelf exercises and content would help first-time users explore the potential, for example in marker-based AR where libraries of printable codes are emerging, as in the case of Google Expeditions (Kitchen, 2021).

To exploit some of the capabilities of high-end sensor-based AR it will need to become scalable across mobile devices (Adams et al., 2013), and to some extent this is happening through the use of machine vision algorithms to recognise elements in a scene. Nevertheless, simplicity of set-up and use will be crucial for broader adoption, especially given the diversity of devices that may be involved, from students' own devices to school-based tablet computers. The same could be said for PARM displays which involve the production of 3D-printed models and the construction of projection rigs. The computing power is modest and the authoring environment can be as simple as PowerPoint, however it still requires a good deal of effort to set up such technologies, and may therefore prove prohibitive for many schools. For those wanting to try this technique, a regular laptop running PowerPoint connected to a portable projector on a camera tripod would provide the basic configuration. The creation of the 3D landscape model would be the most complex element but this could be achieved through collaboration with design and technology teachers with access to 3D printers, perhaps initially exploring some off-the-shelf printable models (for example thingiverse[4]).

Location-based triggering of content is a technically easier solution for augmenting landscape than full sensor-based AR but relies on the design of appropriate trigger zones to avoid an unsatisfactory user experience (Randell et al., 2006). An approach that may help to acknowledge these issues, as described in case study 1, is to gamify the process once a user enters a trigger zone, for example by suggesting that there is now something to search for, or to ask the user to move to the view shown. So whilst authoring solutions may become more widely available, design guidelines will be required for those creating the content to ensure the overall experience functions as intended.

Conclusion

AR in the broadest sense of the term offers exciting possibilities for the creation of learning resources which combine the benefits of multimedia learning with a kinaesthetic mode of delivery, both in the classroom and in the field. There are some interesting design challenges to be aware of if implementing such techniques to ensure that the virtual content complements the real-world experience in ways that support geographical learning rather than detract from it. That said, from a geographical perspective, because AR can promote learning through exploration, and harness the power of digital multimedia in engaging ways, it has great potential for geographical education across all phases.

Notes

1 www.apple.com/uk/ios/augmented-reality/
2 https://developers.google.com/ar/
3 https://arsandbox.ucdavis.edu/
4 www.thingiverse.com/Field_Studies_Council/designs

Acknowledgements

The mediascape featured in Figure 14.1 was developed by Gemma Polmear, working with Richard Poole at Clifton School, Nottingham. This formed part of the Spatial Literacy in Teaching (SPLINT) project, a HEFCE-funded Centre for Excellence in Teaching and Learning. The development of the PARM display featured in Figure 14.2 was undertaken in collaboration with JBA Trust using models of flood events created by JBA Consulting. Many thanks to the staff and students of Keswick School, Cumbria, in particular geography teachers Sue Richardson, Bev Smith and Simon Purdy, for their enthusiastic exploration of the possibilities of PARM in the classroom.

References

Adams, A., FitzGerald, E. and Priestnall, G., 2013. Of catwalk technologies and boundary creatures. *ACM Transactions of Computer-Human Interaction*, 20 (3), Article 15.

Azuma, R.T., 1997. A survey of augmented reality. *Presence-Teleoperators and Virtual Environments*, 6 (4), pp.355–385.

Bartie, P.J. and Mackaness, W.A., 2006. Development of a speech-based augmented reality system to support exploration of cityscape. *Transactions in GIS*, 10 (1), pp.63–86.

Benford, S., Koleva, B., Quinn, A., Thorn, E., Glover, K., Preston, W, Hazzard, A., Rennick-Egglestone, S., Greenhalgh, C. and Mortier, R., 2017. Crafting interactive decoration. *ACM Transactions of Computer-Human Interaction*, 24 (4), Article 26.

Bimber, O. and Raskar, R., 2005. Spatial augmented reality: A modern approach to augmented reality. In *Proceedings of Annual Conference on Computer Graphics and Interactive Techniques – SIGGRAPH'05*. New York: Association of Computing Machinery.

Bower, M., Howe, C., McCredie, N., Robinson, A. and Grover, D., 2014. Augmented reality in education: Cases, places and potentials. *Educational Media International*, 51 (1), pp.1–15.

Bursztyn, N., Walker, A., Shelton, B. and Pederson, J., 2017. Increasing undergraduate interest to learn geoscience with GPS-based augmented reality field trips on students' own smartphones. *GSA Today*, 27 (5), pp.4–11.

Carbonell, C., Saorín Pérez, J.L. and De la Torre Cantero, J., 2018. Teaching with AR as a tool for relief visualization: usability and motivation study. *International Research in Geographical and Environmental Education*, 27 (1), pp.69–84.

Carswell, J.D., Gardiner, K. and Yin, J., 2010. Mobile visibility querying for LBS. *Transactions in GIS*, 14 (6), pp.791–809.

Chen, Z., Wang, Y., Sun, T., Gao, X., Chen, W., Pan, Z., Qu, H. and Wu, Y., 2017. Exploring the design space of immersive urban analytics. *Visual Informatics*, 1, pp.132–142.

Clini, P., Frontoni, E., Quattrini, R. and Pierdicca, R., 2014. Augmented reality experience: From high-resolution acquisition to real time augmented contents. *Advances in Multimedia*, 2014 (18). doi: doi:10.1155/2014/597476.

Colley, A., Thebault-Spieker, J., Yilun Lin, A., Degraen, D., Fischman, B., Häkkilä, J., Kuehl, K., Nisi, V., Jardim Nunes, N., Wenig, N., Wenig, D., Hecht, B., Schöning, J., 2017. The geography of Pokémon GO: Beneficial and problematic effects on places and movement. In *Proceedings of the SIGCHI Conference on Human Factors in Computing Systems. Denver, Colorado, USA, May 2017*. New York: Association for Computing Machinery.

Dourish, P., 2006. Re-spac-ing place: 'Place' and 'space' ten years on. In *Proceedings of the 2006 ACM Conference on Computer Supported Cooperative Work, 2006 (CSCW '06), Banff, AB, Canada*. New York: Association for Computing Machinery.

Dykes, J., Moore, K. and Wood, J., 1999. Virtual environments for student fieldwork using networked components. *International Journal of Geographical Information Science*, 13 (4), pp.397–416.

Efstathiou, I., Kyza, E.A. and Georgiou, Y., 2018. An inquiry based augmented reality mobile learning approach to fostering primary school students' historical reasoning in non-formal settings. *Interactive Learning Environments*, 26 (1), pp.22–41.

Facer, K., Joiner, R., Stanton, D., Reid, J., Hull, R. and Kirk, D., 2004. Savannah: Mobile gaming and learning? *Journal of Computer Assisted Learning*, 20 (6), pp.399–409.

Gill, L. and Lange, E., 2015. Getting virtual 3D landscapes out of the lab. *Computers, Environment and Urban Systems*, 54, pp.356–362.

Hwang, G.J. and Wu, P.H., 2014. Applications, impacts and trends of mobile technology-enhanced learning: A review of 2008–2012 publications in selected SSCI journals. *International Journal of Mobile Learning and Organisation*, 8 (2), pp.83–95.

Hwang, G.J., Po-Han, W., Chi-Chang, C. and Nien-Ting, T., 2016. Effects of an augmented reality-based educational game on students' learning achievements and attitudes in real-world observations. *Interactive Learning Environments*, 24 (8), pp.1895–1906.

Hwang G.J., Wu, P.H. and Ke, R.H., 2011. An interactive concept map approach to supporting mobile learning activities for natural science courses. *Computers and Education*, 57, pp.2272–2280.

Jamali, S.S., Shiratuddin, M.F. and Wong, K., 2014. An overview of mobile augmented reality in higher education. *International Journal on Recent Trends in Engineering and Technology*, 11 (1), pp.229–238.

Kamarainen, A.M., Metcalf, S., Grotzer, T., Browne, A, Mazzuca, D., Tutwiler, M.S. and Dede, C., 2013. EcoMOBILE: Integrating augmented reality and probeware with environmental education field trips. *Computers and Education*, 68, pp.545–556.

Kitchen, B., 2021. Using mobile virtual reality to enhance fieldwork experiences in school geography. In N. Walshe and G. Healy (eds), *Geography Education in the Digital World*. London: Routledge, pp.131–141.

Liao, T. and Humphreys, L., 2015. Layar-ed places: Using mobile augmented reality to tactically reengage, reproduce, and reappropriate public space. *New Media and Society*, 17 (9), pp.1418–1435.

Lu, S.J. and Liu, Y.C., 2015. Integrating augmented reality technology to enhance children's learning in marine education. *Environmental Education Research*, 21 (4), pp.525–541.

McCauley, D.J., 2017. Digital nature: Are fieldtrips a thing of the past? *Science*, 358 (6361), pp.298–300.

McMorrow, J., 2005. Using a web-based resource to prepare students for fieldwork: Evaluating the Dark Peak virtual tour. *Journal of Geography in Higher Education*, 29 (2), pp.223–240.

Meek, S., Priestnall, G. and Goulding, J., 2013. Mobile capture of remote points of interest using line of sight modeling. *Computers and Geosciences*, 52, pp.334–344.

Milgram, P. and Kishino, F.A., 1994. Taxonomy of mixed reality visual displays. *IECE Transactions on Information and Systems* (Special Issue on Networked Reality), E77-D (12), pp.1321–1329.

Mitsova, H., Mitas, L., Ratti, C., Ishii, H., Alonso, J. and Harmon, R., 2006. Real-time landscape model interaction using a tangible geospatial modelling environment. *IEEE Computer Graphics and Applications*, 26 (4), pp.55–63.

Moreno, R. and Mayer, R.E., 1999. Cognitive principles of multimedia learning: The role of modality and contiguity. *Journal of Educational Psychology*, 91 (2), pp.358–368.

Piekarski, W. and Thomas, B.H., 2003. Augmented reality user interfaces and techniques for outdoor modelling. In *Proceedings ACM SIGGRAPH 2003 Symposium on Interactive 3D Graphics. Monterey, CA, USA, April 2003*. New York: Association for Computing Machinery.

Piper, B., Ratti. C. and Ishii H., 2002. Illuminating clay: A 3D tangible interface for landscape analysis. In *Proceedings of the Conference on Human Factors in Computing Systems (CHI '02). Minneapolis, MN, USA, April 2002*. New York: Association for Computing Machinery.

Priestnall, G., 2009. Landscape visualization in fieldwork. *Journal of Geography in Higher Education*, 33 (1), pp.104–112.

Priestnall, G., Fitzgerald, E., Meek, S., Sharples, M. and Polmear, G., 2019. Augmenting the landscape scene: Students as participatory evaluators of mobile geospatial technologies. *Journal of Geography in Higher Education*, 43 (2), pp.131–154.

Priestnall, G., Gardiner, J., Durrant, J. and Goulding, J., 2012. Projection augmented relief models (PARM): Tangible displays for geographic information. In *Proceedings from Electronic Visualisation and the Arts (EVA), London, 2012*. Berlin, Heidelberg: Springer.

Randell, C., Geelhoed, E., Dix, A. and Muller, H., 2006. Exploring the effects of target location size and position system accuracy on location based applications. In K.P. Fishkin, B. Schiele, P. Nixon and A. Quigley (eds), *Pervasive Computing: Pervasive 2006* (LNCS, Vol. 3968). Berlin, Heidelberg: Springer, pp.305–320.

Rogers, Y., Price, S., Fitzpatrick, G., Fleck, R., Harris, E., Smith, H., Randell, C., Muller, H., O'Malley, C., Stanton, D., Thompson, M. and Weal, M.J., 2004. Ambient wood: Designing new forms of digital augmentation for learning outdoors. In *Third International Conference for Interaction Design and Children (IDC 2004). College Park, MD, June 2004*. New York: Association for Computing Machinery.

Sommerauer, P. and Müller, O., 2014. Augmented reality in informal learning environments: A field experiment in a mathematics exhibition. *Computers and Education*, 79, pp.59–68.

Squire, K. and Klopfer, E., 2007. Augmented reality simulations on handheld computers. *The Journal of Learning Sciences*, 16 (3), pp.371–413.

Stenton, S.P., Hull, R., Goddi, P.M., Reid, J.E., Clayton, B.J., Melamed, T.J. and Wee, S., 2007. Mediascapes: Context-aware multimedia experiences. *IEEE Multimedia*, 14 (3), pp.98–105.

Vlahakis, V., Ioannidis, N., Karigiannis, J., Tsotros, M. and Gounaris, M., 2002. Virtual reality and information technology for archaeological site promotion. In *Proceedings of the 5th International Conference on Business Information Systems. Poznań, Poland, 24–25 April 2002*. Poznań: University of Economics, Poznań.

Woods, T.L., Reed, S., His, S., Woods, J.A. and Woods, M.R., 2016. Pilot study using the augmented reality sandbox to teach topographic maps and surficial processes in introductory geology labs. *Journal of Geoscience Education*, 64, pp.199–214.

Chapter 15

Location-based games for geography and environmental education

Steffen Schaal

Introduction

Mobile location-based learning and game-related approaches are increasingly becoming part of educational practices, showing an impact on students' achievement, emotional factors and subject interest while learning (e.g. Lai et al., 2007; Avouris and Yiannoutsou, 2012; Kamarainen et al., 2013; Schaal et al., 2017; Schaal et al., 2018; So and Seo, 2018; Huizinga et al., 2019). The attractiveness of location-based games (LBG) in general has been widely known since blockbuster games such as PokémonGo[1], Ingress[2] or Zombies, Run![3] were released, and educators have started to consider immersion or game-related enjoyment more frequently in the process of learning outside the classroom. Based on a situated cognition framework (Brown et al., 1989), location- or place-based learning could be a way to address important global challenges, starting at the students' doorsteps through 'social, humanistic, and scientific engagement with their surroundings' (Gosslin et al., 2016, p.532), and providing exemplary insights into geoscience key concepts and challenges (Brown et al., 1989). Combining game-based and location-based learning creates what is known as a 'geogame' (Ahlqvist and Schlieder, 2018), requiring locomotion and activities in real places. Geogames provide a link between formal learning processes, inquiry-based learning at distinct locations and playing as 'an active form of entertainment' (Schaal et al., 2018, p.218). The design of such an educational LBG is challenging, and requires:

- Relevant place-specific content knowledge.
- An understanding of spatial game mechanics.
- Pedagogical content knowledge to select and provide adequate location-based tasks.
- Considerable digital skills.
- Time 'for creating, testing and improving the design' (Schlieder et al., 2018, p.112).

The potential of LBG is not just in playing them, but also in their construction by learners (Klopfer and Sheldon, 2010; Schaal and Lude, 2015; Vogel and Perry, 2018), especially from a perspective of spatial thinking in the geography classroom (Hespanha et al., 2007; Kuhn, 2012; Kerski, 2015; Yang and Chang, 2017; Schlieder et al., 2018). With this in mind, this chapter gives an overview of the evolution of location-based games for educational purposes, provides empirical insight into evidence for using location-based games to support learning in several subjects, and finally discusses future trends for location-based games.

From location-based games to educational geogames

Scavenger hunts and other hide-and-seek games are undoubtedly popular, and they are used all over the world in out-of-school settings like museums, botanical gardens and natural sites. Location-based games can bring a lot of enjoyment combining cognitive challenges when solving riddles, mostly in collaboration with other players, and the joy of locomotion related to the physical world (Vorderer et al., 2004). Placing a game experience in the real world instead of playing a traditional board game reflects the premise that individuals interact with the physical world, cognitively as well as emotionally, they experience it holistically, they interact with others and thus a space becomes a personally relevant place (Harrison and Dourish, 1996; Lengen, 2016; Olesky and Wnuk, 2017). Digital LBGs can be seen as derivatives of geocaching (Schaal and Lude, 2015; Schlieder et al., 2018), which uses GPS-coordinates and receivers to find hidden caches in the environment (McNamara, 2004). Combining several single caches results in a so-called multi-cache: players have to find a cache and solve a riddle or a task to get the next GPS-coordinates or receive a hint to the next cache. This leads to a kind of sequential treasure hunt. As the functionality of GPS-receivers was limited in the early 2000s, further information had to be provided physically at the caches, or a player had to download all relevant information from a website (e.g. geocaching.com) in advance.

The next step of LBG evolution was so-called mixed-reality or pervasive games with more elaborate game designs (Hinske et al., 2007). Compared to the real-world experiences of geocaching, pervasive games embed more sophisticated game mechanics and add (virtual) components using mobile computing technology. This allows, for instance, for more interactive game formats such as Can You See Me Now?[4] or Zombies, Run! In these examples the players interact not only with the environment but also with a game server or other real players – they have to catch something or avoid being caught (chase and catch game mechanic).

Geogames can be considered as 'all games and play that use geocontent and [are] mediated by GI [geoinformation] technology' (Ahlqvist and Schlieder, 2018, p.1) and therefore they include in a broader sense all LBG genres described above. Ahlqvist and Schlieder (2018) describe the characteristics of GIS/spatial thinking and of game patterns, resulting in an elaborated understanding of geogames/geoplay. In general, geogames use the features of smartphones for GPS-supported orientation/navigation and all information is provided digitally onscreen when it is needed (e.g. text, images, audio or video according to the players' position via GPS coordinates, near-field communication (NFC), beacons, QR code, etc.). Furthermore the integrated camera, microphone, texting, onboard sensors and video features can be used in the game for documentation or for observations (Holloway and Mahan, 2012). Context or location awareness is provided by visual clues (e.g. QR codes), by georeferences (e.g. GPS signal or NFC) or by a combination of both (e.g. find a place and if a predefined radius is entered a location-based information or task is triggered). Geogames can include learning activities (Giannakas et al., 2018) such as:

- Observations, identifications (e.g. real objects, learning environments) and data collection: players, for instance, are navigated to a place where specific observations have to be conducted which are used to solve a location-based task.
- Location-based learning with online support/guidance: information about a place is provided digitally on a mobile device immediately when needed. This could be audio

information, a video or some mixed-reality objects which anchor virtual objects to the real world.
- Problem-solving and decision-making (e.g. via experiments, simulations): within a geogame about urban development, players discover places, they use a simulation to change the site (e.g. increase the size of natural city sites and parks), and they then see the results of their manipulations in a similar way to videogames such as SimCity. Hence the learners discover a site, they learn about urban planning, they discuss possibilities, they test them virtually and then immediately get a result as an outcome of a simulation (e.g. Schneider et al., 2019).

The intentions of designing and playing geogames vary broadly from creating leisure activities to the provision of elaborate educational games across multiple disciplines. Educational geogames have been used to provide experiences across the disciplines of history (e.g. Huizinga et al., 2009; Vassilakis et al., 2017), ecology and environmental education (Schaal and Lude, 2015; Kamarainen et al., 2018; Schneider and Schaal, 2018), language learning (Pitura and Terlecka-Pacut, 2018), biology and biodiversity (Crawford et al., 2017; Schaal et al., 2017; Pernaa, 2018; Schaal et al., 2018), and geography (Kerski, 2015; Feulner, 2016; Pánek et al., 2017).

Geogames and educational outcomes

In educational geogames learners navigate to places where information and tasks are provided. The purpose of playing geogames could be:

- A cognitive achievement related to the place or the general topic of the game (e.g. a specific aspect of the physical geography of a historical site, event or ecosystem).
- An emotional state during playing (e.g. game enjoyment, excitement, immersion).
- An increase of awareness (e.g. perception of biodiversity, spatial awareness).
- A change in attitudes (e.g. towards nature conservation, cultural diversity, etc.).

In the last decade a vast number of studies were conducted in the field of mobile location-based learning and thus empirical evidence is manifold. The educational benefit of location-based learning and geogames is often discussed. For example, Giannakas et al. (2018) highlight the aquisition of context-/location-aware knowledge in their review of mobile game-based learning between 2004 and 2016. The authors assume that 'finding the juste milieu between delightful play and learning outcomes with reference to learning theories' (Giannakas et al., 2018, p.379) would be one of the issues to be addressed in the design and the research of educational geogames. If location-based learning and geogames are well designed, cognitive achievement (e.g. factual and conceptual knowledge) is at least comparable or superior to other field-based activities. For instance, Huizinga et al. (2009) used a geogame to learn about medieval Amsterdam and compared it to a regular, project-based learning setting, finding a higher cognitive achievement (knowledge about medieval Amsterdam) for the learners playing the game. Ruchter et al. (2010) also report at least similar learning outcomes of environmental knowledge exploring a national park by using mobile electronic devices (MEDs) compared to using a traditional guide. In a similar setting, Lai et al. (2007) measured a higher cognitive achievement (biological knowledge and self-generated observations) in students exploring a school garden supported by MEDs, and

Kamarainen et al. (2013) describe teachers' perceptions that using MEDs results in a deeper understanding of scientific work compared to previous field-trips which did not incorporate the use of MEDs. The authors highlight that MEDs allow a more self-determined scientific experience on a fieldtrip and, related to that, more intensive student collaboration in a mostly authentic setting. Furthermore, using augmented reality (AR) facilitates the transfer of what students have already learnt in the classroom to a real-world situation. Most of the studies find that stronger individual involvement within the object-related learning activity results in higher cognitive activation; this is, they suggest, attributable to the more self-directed learning environment of geogames, as well as their collaborative nature as they are often played in small groups.

If awareness and attitudes are addressed within a geogame, elaborate research designs are required and the empirical findings become more complex. Evidence that playing specific geogames leads to a related change in attitudes is reported in several studies (Lai et al., 2007; Ruchter et al., 2010; Crawford et al., 2017), but often without respecting the reciprocity of factors like interest, game-related enjoyment, prior knowledge or other player characteristics. Schaal et al. (2018) analysed cognitive achievement and attitudes towards nature in a location-based simulation game. They analysed the achievement of biodiversity-related knowledge and they found that it was neither influenced by previous attitudes nor by their enjoyment during the gameplay. More interestingly, they found that game-related enjoyment was a strong predictor for the player's attitudes towards nature and, combined with other players' requisites (e.g. general ecological behaviour as an almost stable individual trait), even a short geogame session appeared to contribute to an attitudinal change. Hence they conclude that focusing just on the assessment of learning outcomes might risk an oversimplification of the complex cognitive, social and emotional processes that occur while playing educational geogames. This is in line with the review of Giannakas et al. (2018) highlighting a six-dimensional framework for the evidence-based design of geogames. The authors focus on specific pedagogies of geogames, including not only technological aspects but also pedagogical dimensions such as the spatio-temporal design, collaboration and social aspects or the need for personalisation (learner profiles, learning content, content representation).

As well as the educational benefits of location-based gameplay, research focuses increasingly on students' learning through the process of designing their own geogames. Klopfer and Sheldon (2010) describe the opportunities of engaging learners in the game design, highlighting the exploration of location-based context, the generation of content and the design of authentic and relevant scenarios. Vogel and Perry (2018) describe a facilitator-youth-apprenticeship approach for the design of geogames, suggesting that collaboration and deeper connections to the learners' local communities were made. Slussareff and Boháčková (2016) compared the outcomes of designing and playing a geogame in the subject of history with students aged 14–15 years in an explorative study. Their results point towards better learning outcomes which the authors relate to deeper content processing while being actively engaged in the game design. Edmonds and Smith (2017) report similar results with learners in higher education and they conclude that 'both the playing and designing of [geogames] by students can provide benefits by delivering active, engaging, and authentic educational experiences, which enhance the opportunities to interact with locations, online content, and with each other' (p.51). Schaal and Baisch (2017) conducted an explorative study with primary school children (aged 10 years) collaboratively designing a geogame focusing on tree botany, and they encountered challenges related to the cognitive load due to the content and the challenges of game design. They suggest increasing student involvement in the design process according to the students' capabilities (e.g. from the selection of an adequate location to the content design and the provision of location-based tasks). In general, while empirical evidence in this field remains scarce, research suggests that educational geogames offer both promise and challenge for geography educators; in particular, technological skills and professional content knowledge have to be combined to achieve success, with several studies highlighting the need for clear design frameworks as well as easy-to-use authoring systems (e.g. Bartsch et al., 2017; Oppermann et al., 2018).

Process of geogame design

Geogame design can be described from various perspectives (e.g. focus on ICT, game mechanics, educational goals, digital storytelling) and is grounded within several disciplines (e.g. biology, computing, geography and history). Ahlqvist and Schlieder (2018) make a comprehensive assessment of geogame/geoplay design patterns and, combined with other findings (Avouris and Yiannoutsou, 2012; Giannakas et al., 2018; Oppermann et al., 2018; Schaal et al., 2018; Ferreira et al., 2019), four major areas for balancing a geogame can be derived (Schaal and Schaal, 2018; Figure 15.1).

The first step of an iterative educational geogame design process should be setting the learning objective, the contents' core concepts and the learning tasks before identifying relevant places. In the context of geography education, for instance, one does not need much imagination to relate geogames for almost all the discipline's key concepts (place, space, environment, interconnection, sustainability, scale and change; Clifford et al., 2008). In the next steps a decision will be needed about the sequencing of the tasks or game mechanics, which depends, again, on the geolocation and the learners' prior knowledge (Figure 15.2).

If, on the one hand, the landscape forces the players to follow a predefined route, not every place of interest will necessarily be an ideal location to undertake a geogame. In addition, when several players simultaneously follow the same route it can be difficult to

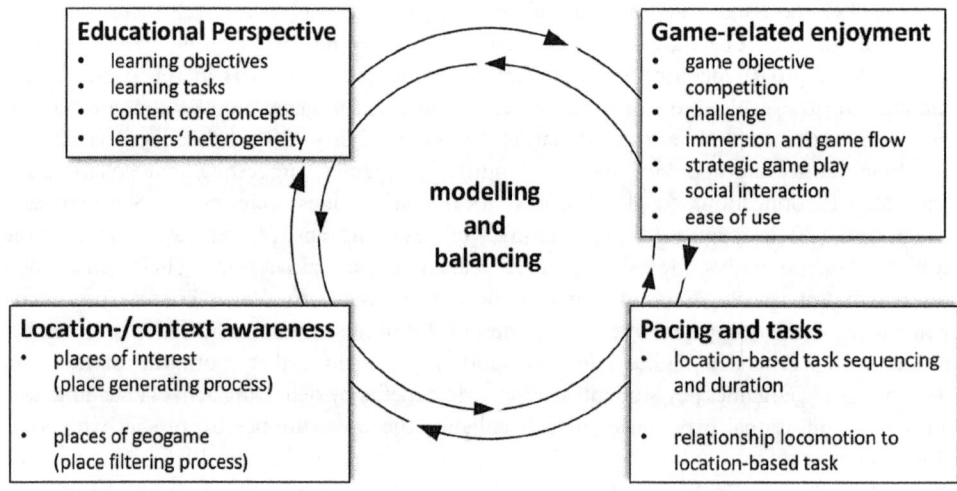

Figure 15.1 Areas for geogame design (adapted from Schaal and Schaal, 2018, p.44).

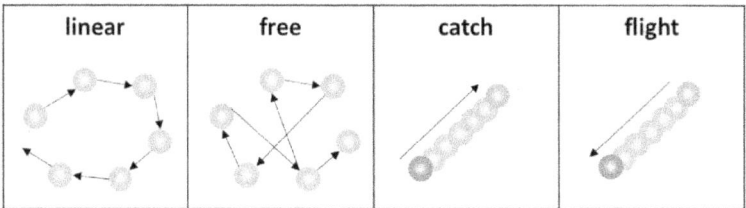

Figure 15.2 Sequencing of location-based tasks.

avoid one player catching up with another which influences the overarching game mechanics. In this case, a catch'n'hunt game will soon become boring because the running speed, and not the strategic decisions a player makes, will determine whether the game objective is achieved. If, on the other hand, the sequencing of the location-based tasks is open, players are able to decide on the strategies used to access every location. This could become a part of the game mechanics and, at the same time, allow for a balance between the proportion of time needed for movement from one place to another, and the time needed for completion of the location-based tasks themselves. In this way, game design needs to consider both game mechanics (moving from one location to another, for example in games such as seek'n'find, hunt'n'catch, or strategic games like TicTacToe), and the plot of the game, both of which affect players' ultimate immersion (Spallazzo and Ilaria, 2018).

In the next iteration, after filtering relevant places for the geogame (including consideration of mobile internet connectivity or GPS signal strength; see Yanenko et al., 2017), an educator has to define to what extent a task is related to the place and how challenging it is. While the task for players in the geogame Tidy City (Wetzel et al., 2011), for instance, is just to find a place using a visual hint (e.g. an architectural artefact) and to solve a riddle there, learners in the BioDiversity2GoSimulation game (Schaal et al., 2017; Schaal et al., 2018) navigate to a location, they conduct a location-based inquiry task and according to the game narrative they then use the results to manipulate a virtual simulation, balancing ecological and economic outcomes to manage a forest sustainably. While the first game needs only a limited time for the location-based tasks (players spend more time moving between tasks), the latter focuses more on the tasks themselves and on deeper learning through the location-based inquiry.

The design of educational geogames is not trivial and, as shown, several factors have to be reflected in the decisions made. This is challenging but it provides a high potential for learning – for pupils, as well as for teachers. Thus, teacher education should endeavour to provide opportunity for reflection on the design of geogames in different subjects, including geography education. Several authoring systems focus on the educational design rather than on technological skills and so almost anyone can design a geogame without knowledge of coding.

Future trends for location-based games

Mobile technology is rapidly evolving and within the next decade the accessibility of powerful mobile internet and geolocation systems will facilitate location-based gaming. This will have an impact both on the ease with which teachers can create geogames and on how educational geogames are played. In particular:

- It remains challenging to ensure a stable mobile internet connection, and game designers have to either select places with reliable network coverage or provide apps running without connectivity. As such, current geogames rarely provide real-time interaction with either an intelligent game server or with other connected players. With improved connectivity, this will change and more complex game mechanics will appear, for instance real-world simulation games where the decisions of players at a location will immediately influence the environment and other players. This will also include player-generated content as a part of the game experience and will provide a mechanism for gamified approaches to citizen science.
- AR applications are already provided for laypersons and it can be expected that this technology will continue to be developed into mobile educational games. It can be expected that geographical sites will provide opportunities for interactive AR; for example, exploring the formation of landscapes through merging real and virtual experiences (see Priestnall (2021) for further discussion of AR).
- Learning analytics will improve the provision of information according to the players' interaction within a location-based task and, therefore, learners' heterogeneity could be addressed more easily in adaptive geogames. For instance, an intelligent adaptive game engine is able to identify the users' abilities within a few game interactions, and subsequently provide a customised aspiration level, appropriate information or support to learners.

The design of educational geogames will evolve:

- Recent authoring systems for the customer market provide georeferencing and the provision of different location-based tasks and information; game mechanics are mostly simple (e.g. treasure hunts like Actionbound,[5] Seek'n'Spot,[6] Elspoto,[7] MILE[8]) or more demanding from a computing perspective (e.g. TaleBlazer[9]). Thus, this will allow every (geography) educator to create a playable geogame.
- Including AR content in location-based tasks is likely to become easier, and educators will be able to design and provide information in a more immersive way. This will make abstract content (e.g. the geology of an area) more accessible.

Conclusion

As shown, the benefits of geogames – as a player or as a designer – are manifold, but it is also challenging to include them in an appropriate way into the teaching and learning of geography. Educators need to reflect carefully and in a multi-perspective way on the usefulness of geogames for location-based geographical learning. Further, they must decide whether media-assisted learning is sufficient to support students with both the acquisition of knowledge and the development of attitudes, or whether an informed, conscious decision should be made to turn off mobile technology and instead immerse them in the real world.

Notes

1 www.pokemongo.com
2 www.ingress.com/
3 https://zombiesrungame.com/
4 www.blasttheory.co.uk/projects/can-you-see-me-now/

5 www.actionbound.com
6 https://seeknspot.fromlabs.com/
7 www.espoto.com
8 www.mile-bw.de
9 www.taleblazer.org

References

Ahlqvist, O. and Schlieder, C., 2018. Introducing geogames and geoplay: Characterizing an emerging research field. In O. Alqvist and C. Schlieder (eds), *Geogames and Geoplay*. Cham: Springer, pp.1–18.

Avouris, N.M. and Yiannoutsou, N., 2012. A review of mobile location-based games for learning across physical and virtual spaces. *Journal of Universal Computer Science*, 15 (18), pp.2120–2142.

Bartsch, S., Müller, H., Oppermann, L. and Schaal, S., 2017. Using smartphones for tracing local food: Location-based games on mobile devices in consumer and nutrition education. In R. Stiftung (ed.), *Places of Food Production. Origin, Identity, Imagination. Proceedings of the 21st International Ethnological Food Research Conference*. Heidelberg: Springer, pp.235–248.

Brown, J.S., Collins, A. and Duguid, P., 1989. Situated cognition and the culture of learning. *Educational Researcher*, 18 (1), pp.32–42.

Clifford, N., Holloway, S., Rice, S. and Valentine, G., 2008. *Key Concepts in Geography*. London: Sage.

Crawford, M., Holder, M. and O'Connor, B., 2017. Using mobile technology to engage children with nature. *Environment and Behavior*, 49 (9), pp.959–984.

Edmonds, R. and Smith, S., 2017. From playing to designing: Enhancing educational experiences with location-based mobile learning games. *Australasian Journal of Educational Technology*, 33 (6), pp.41–53.

Ferreira, C., Maia, L.F., de Salles, C., Trinta, F. and Viana, W., 2019. Modelling and transposition of location-based games. *Entertainment Computing*, 30, 100295.

Feulner, B., 2016. Geogames in geography education: A design-based research study. In *Proceedings of the Workshop on Geogames and Geoplay. AGILE 2016. Helsinki, June 14–17, 2016*. Available at www.geogames-team.org/agile2016/submissions/ Feulner_Geography_Education.pdf [Accessed 1 September 2019].

Giannakas, F., Kambourakis, G., Papasalouros, A. and Gritzalis, S., 2018. A critical review of 13 years of mobile game-based learning. *Educational Technology Research and Development*, 66 (2), pp.341–384.

Gosslin, D., Burian, S., Lutz, T. and Maxson, J., 2016. Integrating geoscience into undergraduate education about environment, society, and sustainability using place-based learning: Three examples. *Journal of Environmental Studies and Sciences*, 6 (3), pp.531–540.

Harrison, S. and Dourish, P., 1996. Re-place-ing space: The roles of place and space in collaborative systems. In *Proceedings ACM Conference on CSCW'96 (Boston, MA)*. New York: ACM.

Hespanha, S., Rebich, S., Goodchild, F. and Janelle, D.G., 2007. Spatial thinking and technologies in the undergraduate social science classroom. *Journal of Geography in Higher Education*, 33 (S1), pp.17–27.

Hinske, S., Lampe, M., Magerkurth, C. and Röcker, C., 2007. Classifying pervasive games: On pervasive computing and mixed reality. In K. Jergers (ed.), *Concepts and Technologies for Pervasive Games: A Reader for Pervasive Gaming Research*. Aachen: Shaker, pp.1–20.

Holloway, P. and Mahan, C., 2012. Enhance nature exploration with technology. *Science Scope*, 35 (9), pp.23–28.

Huizinga, J., Admiraal, W., Akkerman, S. and ten Dam, G., 2009. Mobile game-based learning in secondary education: Engagement, motivation and learning in a mobile city game. *Journal of Computer Assisted Learning*, 25 (4), pp.332–344.

Huizinga, J., Admiraal, W., ten Dam, G. and Voogt, J., 2019. Mobile game-based learning in secondary education: Students' immersion, game activities, team performance and learning outcomes. *Computers in Human Behavior*, 99, pp.137–143.

Kamarainen, A.M., Metcalf, S., Grotzer, T., Browne, A., Mazzuca, D., Tutwiler, M.S. and Dede, C., 2013. EcoMOBILE: Integrating augmented reality and probeware with environmental education field trips. *Computers & Education*, 68, pp.545–556.

Kamarainen, A.M., Reilly, J., Metcalf, S., Grotzer, T. and Dede, C., 2018. Using mobile location-based augmented reality to support outdoor learning in undergraduate ecology and environmental science courses. *Bulletin of the Ecological Society of America*, 99 (2), pp.259–276.

Kerski, J., 2015. Geo-awareness, geo-enablement, geotechnologies, citizen science, and storytelling: Geography on the world stage. *Geography Compass*, 9 (1), pp.14–26.

Klopfer, E. and Sheldon, J., 2010. Augmenting your own reality: Student authoring of science-based augmented reality games. *New Directions for Youth Development*, 128, pp.85–94.

Kuhn, W., 2012. Core concepts of spatial information for transdisciplinary research. *International Journal of Geographical Information Science*, 26 (12), pp.2267–2276.

Lai, C.H., Yang, J.C., Chen, F.C., Ho, C.W. and Chan, T.W., 2007. Affordances of mobile technologies for experiential learning: The interplay of technology and pedagogical practices. *Journal of Computer Assisted Learning*, 23 (4), pp.326–337.

Lengen, C., 2016. Places: Orte mit Bedeutung. In U. Gebhard and T. Kistemann (eds), *Landschaft, Identität und Gesundheit*. Wiesbaden: Springer, pp.19–29.

McNamara, J., 2004. *Geocaching for Dummies*. New York: John Wiley & Sons.

Olesky, T. and Wnuk, A., 2017. Catch them all and increase your place attachment! The role of location-based augmented reality games in changing people-place relations. *Computers in Human Behavior*, 76, pp.3–8.

Oppermann, L., Schaal, S., Eisenhardt, M., Brosda, C., Müller, H. and Bartsch, S., 2018. Move, interact, learn, eat: A toolbox for educational location-based games. In *Advances in Computer Entertainment Technology. ACE 2017. Lecture Notes in Computer Science*. Cham: Springer.

Pánek, J., Gekker, A., Hind, S., Wendler, J., Perkins, C. and Lammes, S., 2017. Encountering place: Mapping and location-based games in interdisciplinary education. *The Cartographic Journal*, 3 (55), pp.285–297.

Pernaa, J., 2018. Digital gaming for evolutionary biology learning: The case study of parasite race, an augmented reality location-based game. *LUMAT: International Journal on Math, Science and Technology Education*, 6 (1), pp.1–25.

Pitura, J. and Terlecka-Pacut, E., 2018. Action research on the application of technology assisted urban gaming in language education in a Polish upper-secondary school. *Computer Assisted Language Learning*, 31 (7), pp.734–763.

Priestnall, G., 2021. Augmented reality: Opportunities and challenges. In N. Walshe and G. Healy (eds), *Geography Education in the Digital World*. London: Routledge, pp.155–167.

Puentedura, R., 2014. *Learning, Technology, and the SAMR Model: Goals, Processes and Practice*. Available at www.hippasus.com/rrpweblog/archives/000127.html [Accessed 1 Septemter 2019].

Ruchter, M., Klar, B. and Geiger, W., 2010. Comparing the effects of mobile computers and traditional approaches in environmental education. *Computers & Education*, 54 (4), pp.1054–1067.

Schaal, S. and Baisch, P., 2017. Natur und Kultur, erspielen: Geogames gestalten mit Schülerinnen und Schülern im Sachunterricht (Projekt 'Na KueG!') [Playing nature and culture: Designing geogames with primary science students]. *Ludwigsburger Beiträge zur Medienpädagogik online* [*Ludwigsburg Contribution to Media Education*], 17, pp.1–12.

Schaal, S. and Lude, A., 2015. Using mobile devices in environmental education and education for sustainable development: Comparing theory and practice in a nation-wide survey. *Sustainability*, 7 (8), pp.10153–10170.

Schaal, S. and Schaal, S., 2018. Entdecke die Natur: Ortsbezogene Geogames entwerfen und anwenden [Discover nature: Design and use of location-based geogames]. *Unterricht Biologie* [*Biology Education*], 433, pp.44–47.

Schaal, S., Otto, S., Schaal, S. and Lude, A., 2018. Game-related enjoyment or personal pre-requisites – which is the crucial factor when using geogames to encourage adolescents to value local biodiversity. *International Journal of Science Education Part B*, 8 (3), pp.213–226.

Schaal, S., Schaal, S. and Lude, A., 2017. BioDiv2Go: Does the location-based geogame 'FindeVielfalt Simulation' increase the valuing of local biodiversity among adolescent players? In N. Gericke and M. Grace (eds), *Challenges in Biology Education Research: A Selection of Papers Presented at the 11th Conference of European Researchers in Didactics of Biology (ERIDOB)*. Karlstad: Karlstads University, pp.315–332.

Schlieder, C., Kremer, D. and Heinz, T., 2018. Teaching geogame design: Game relocation as a spatial analysis task. In C. Schlieder and O. Ahlqvist (eds), *Geogames and Geoplay*. Cham: Springer, pp.111–130.

Schneider, J. and Schaal, S., 2018. Location-based smartphone games in the context of environmental education and education for sustainable development. Fostering connectedness to nature with Geogames. *Environmental Education Research*, 24 (11), pp.1597–1610.

Schneider, J., Schaal, S. and Schlieder, C., 2019. Integrating simulation tasks into an outdoor location-based game flow. *Multimedia Tools and Applications*. doi: doi:10.1007/s11042-019-07931-4.

Slussareff, M. and Boháčková, P., 2016. Students as game designers vs 'just' players: Comparison of two different approaches to location-based games implementation into school curricula. *Digital Education Review*, 29, pp.284–297.

So, H.-J. and Seo, M., 2018. A systematic literature review of game-based learning and gamification research in Asia. In K.J. Kennedy and J.C. Lee (eds), *Routledge International Handbook of Schools and Schooling in Asia*. London: Routledge, pp.396–410.

Spallazzo, D. and Ilaria, M., 2018. Stories, metaphors and disclosures: A narrative perspective between interaction and agency. In C. Schlieder and O. Ahlqvist (eds), *Location-Based Mobile Games*. Cham: Springer, pp.59–82.

Vassilakis, K., Charalampakos, O., Glykokokalos, G., Kontokalou, P., Kalogiannakis, M. and Vidakis, N. 2017. Learning history through location-based games: The fortification gates of the Venetian walls of the city of Heraklion. In A.L. Brooks, E. Brooks and C. Sylla (eds), 2017. *Interactivity, Game Creation, Design, Learning, and Innovation*. Cham: Springer, pp.510–519.

Vogel, S. and Perry, J., 2018. We got this: Toward a facilitator-youth apprenticeship approach supporting collaboration and design challenges in youth-designed mobile location-based games. In S. Arafeh, D. Herro, C. Holden and R. Ling (eds), *Mobile Technologies: Perspectives on Policy and Practice*. Charlotte, NC: IAP, pp.143–168.

Vorderer, P., Klimmt, C. and Ritterfeld, U., 2004. Enjoyment: At the heart of media entertainment. *Communication Theory*, 14 (4), pp.388–408.

Wetzel, R., Blum, L., Feng, F., Oppermann, L. and Straeubig, M., 2011. Tidy city: A location-based game for city exploration based on user-created content. In M. Eibl (eds), *Mensch & Computer 2011: überMEDIEN|ÜBERmorgen*. München: Oldenbourg Verlag, pp.487–496.

Yanenko, O., Stein, K. and Klug, C., 2017. *Challenges in Geogame Design for Biodiversity Education*. Available at: www.geogames-team.org/agile2017/submissions/Biodiversity%20Education_AGILE_2017.pdf [Accessed 1 September 2019].

Yang, H.-C. and Chang, W.-C., 2017. Ubiquitous smartphone platform for K-7 students learning geography in Taiwan. *Multimedia Tools and Applications*, 76 (9), pp.11651–11668.

Part V

Conclusion

Chapter 16

From the digital world to the post-digital world

The future generation of geographers

Grace Healy and Nicola Walshe

In 2015, Kerski extolled that 'opportunities in using geospatial technologies as a meaningful and sustained part of education exist as never before' (p.183); he argued that this is not just through technological enhancements, but also the development of 'inquiry, critical thinking, outdoor education, authentic assessment, STEM, technical, green, and other careers, and meaningful teaching with technology' (p.183). While Kerski (2015) was reflecting on the opportunities afforded by geospatial technologies in particular, this illustrates the way that technology has been framed by some as a catch-all for geography teachers, a potential solution for any pedagogical challenge they face. Authors in this book have reflected more critically on the potential of geography education in the digital world, through exploring its influences on teachers' professional practice, the changing way that geographical sources can be sourced, selected and interrogated, and the use of a range of technologies in classroom and field-based contexts. This chapter offers some final comments on the prominent themes, and considers 'What next?' for geography education.

Geography education in the digital world

Woven through many chapters within this book is the notion of teachers' curriculum making, whereby the geography curriculum experienced by students draws on the three sources of energy – *school geography, student experiences*, and *teacher choices* – which are encompassed within the context of the discipline of geography (Lambert and Biddulph, 2015). This approach ultimately places 'ownership of the "local" curriculum in the hands of each geography teacher' (Brooks, 2016, p.77) and yet, in an ever-evolving digital world, these sources of energy are in perpetual motion. As such, it seems pertinent to question how the 'radically dynamic' (Taylor, 2013, p.810) nature of what teachers and students bring to the classroom, both individually and in combination, might be shaped by the digital world.

Taylor (2013) draws on Massey's formation of space as 'the sphere of a multiplicity of trajectories' (Massey, 2005, p.119) to explicate that both teachers and students are not the same people each time they meet in the geography classroom; this is because their 'trajectories [are] different between points of meeting in terms of experiences at home and in school, which media they had engaged with, and relationships and conversations renewed or broken off' (Taylor, 2013, p.811). In the digital world, the interactions between teacher and student are now no longer found only within the walls of the geography classroom, but might take place in different spaces within the digital world through online education platforms such as Google Classroom. In this book, we have seen how both children's geographies and teachers' professional identities are changing in the digital world (Brooks,

2021; Hammond, 2021; Puttick, 2021), yet there remains significant scope to fully understand how the digital world is shaping *student experiences* and how this intersects with the changing identity of teachers and their professional practice. As geographers, we should naturally appreciate the importance of spatial and temporal context, both in shaping the professional landscape for geography teachers at a national, school and department scale (Clandinin and Connelly, 1995; Alexander, 2009), and in shaping students' experiences of the digital world per se. Therefore, we propose that it is even more important that geography teachers are empowered as professionals to navigate what is right to support their own curriculum making within the context of their specific classroom and with their children and resources in mind (Lambert, 2015).

For as long as technology has been introduced into the classroom, from interactive whiteboards to VHS video players, there has been debate as to its geographical purpose and how this might inform *teacher choices* (Brooks, 2018). Within the chapters of this book authors grapple with ways in which technology can enhance geographical education, as it opens up new possibilities for teachers' professional learning (Walshe et al., 2021), provides epistemic access to geographical knowledge (Fargher and Healy, 2021), enables young people as active citizens (Pike, 2021), and develops opportunities for fieldwork (Kitchen, 2021). However, despite this educational potential, Schaal (2021) adds the caveat that, as geography educators, we must remain cognisant that we have the power to immerse our children and young people in the real world *without technology*. Further, if, as Roberts (2021) highlights, this is not just about geography education *in* the digital world, but geography education *for* the digital world, then we must consider the role teachers play in helping students navigate this digital world through their geographical education. This has significant implications for the professional development of geography teachers; for example, it now needs to empower teachers as professionals to be able not only to use technology to develop geographical learning, but provide them with the confidence to know when not to, based on the curricular purpose at hand, or to engage critically with information presented within these 'post-truth times' (Roberts, 2021). Even if initial teacher education responds with the enthusiasm required to better prepare beginning teachers, what of the subsequent potential change in the power dynamic between existing (relatively 'unprepared') and trainee teachers, something that has been identified as an issue previously by Walshe (2017) with regards to GIS expertise.

Whilst the disciplinary resource of geography is itself dynamic in nature (Lambert and Walshe, 2018), the source of energy of *school geography* is mediated by other influences, such as examination specifications (Puttick, 2015). Therefore, to understand this source of energy, there needs to be greater attention to how geospatial technology and digital geographies can be represented in future iterations of curriculum documents and examination specifications (Puttick, 2021), how this curriculum entitlement might vary globally, and how it is translated into international statements, such as the IGU-CGE Charter on Geographical Education (IGU-CGE, 2016). The importance of this curriculum entitlement to geospatial technologies is impressed upon us by Muniz Solari et al. (2015), who argue that 'geography today cannot study geographical issues and resolve them without an effective use of geospatial technologies' (p.v). As with any curriculum reform, there needs to be due consideration to teachers' subject expertise (Brooks, 2016) and how this develops through both initial teacher education and continued professional development, as teachers' recontextualisation of the curriculum will shape geography lessons both in and beyond the classroom.

In the 'context of the discipline of geography' within the digital world, there appears to be greater opportunity to respond to Butt's (2020) call to cultivate connections between the research of geographers and geography educationalists (Butt, 2020); cross-fertilisation between school and academic geography, drawing on the ideas of both geography and education (Lambert, 2015), has the potential to impact on all sources of curriculum making. Firstly, greater engagement with children's geographies could inform how teachers draw on *student experiences*, whilst digital geographies could become more accessible to teachers as a part of the *school geography* through future curriculum reform. There is also value in teachers of geography, in both school and the academy, being able to contribute to and learn from research across their two contexts, to inform *teacher choices*. In this book, Priestnall's (2021) use of augmented reality (AR) crosses boundaries from practice within HE contexts into school classrooms; but if schools engage with the use of AR to support their geography, what might the implications for HE be? The same question might already be asked in relation to geospatial technologies where developments in school-level practice have necessitated consideration of progression and continuity between secondary and tertiary phases. In this way, as school curricula respond to the digital world, there are pedagogical implications for university geographers. While professional relationships between school and university geography do happen, they are often piecemeal and frequently hierarchical (foregrounding expertise of geography academics). We question whether the digital world will provide a more urgent need for a more egalitarian partnership.

The future of geography education in the post-digital world

In thinking about 'What next?' for geography education, we turn to the discourse beyond geography education where there is discussion about the emergence of a post-digital world. The notion of the post-digital involves the merging of worlds, as 'the disjunction between what is digital and what is not is blurred' (Fazi, 2016). At present, the separation 'between digital and non-digital, between virtual and "real"' (Tesar and Hood, 2019, p.103) is perceived to be problematic. Within such a world, the digital is becoming inextricably integrated within teachers' and students' everyday lives (Feenberg, 2019). Yet, it is also undeniable that future generations of children and young people learning geography will only know first-hand this post-digital world. Tesar and Hood (2019) argue that these future generations' world will be 'punctured and constrained by the boundaries imposed by others, who have known the [pre] digital age', whilst these are the generations that will be left with the 'profound implications of the Anthropocene on a grand scale, grappling with the policies and [in]decisions of previous generations' (p.103). This raises significant questions about the extent to which geography teacher education prepares teachers to understand their students' rapidly changing experiences of the world; better understanding children's changing geographies, as proposed by Hammond (2021), may be one avenue that is worth exploring. However, as we progress towards the post-digital age within the next decade and beyond, whilst paying heed to 'who' geography teachers are teaching, we must also not lose sight of 'what' geography education is for (Lambert and Solem, 2017). Within this human epoch of the Anthropocene (Morgan, 2019), there are ever more geographical and educational challenges; as such, 'the enduring and significant tasks of deepening and extending professional repertoires of thought and practice' (Lambert, 2018, p.368) for geography education could not be any more vital.

References

Alexander, R., 2009. *Culture and Pedagogy: International Comparisons in Primary Education.* Oxford: Blackwell.

Brooks, C., 2016. *Teacher Subject Identity in Professional Practice: Teaching with a Professional Compass.* Abingdon: Routledge.

Brooks, C., 2018. Insights on the field of geography education from a review of Masters level practitioner research. *International Research in Geographical and Environmental Education*, 27 (1), pp.5–23.

Brooks, C., 2021. Teacher identity, professional practice and online social spaces. In N. Walshe and G. Healy (eds), *Geography Education in the Digital World.* London: Routledge, pp.7–16.

Butt, G., 2020. *Geography Education Research in the UK: Retrospect and Prospect.* Cham: Springer.

Clandinin, D.J. and Connelly, F.M., 1995. *Teachers' Professional Knowledge Landscapes.* New York: Teachers College, Columbia University.

Fargher, M. and Healy, G., 2021. Empowering geography teachers and students with geographical knowledge: Epistemic access through GIS. In N. Walshe and G. Healy (eds), *Geography Education in the Digital World.* London: Routledge, pp.102–116.

Fazi, M.B., 2016. Review of David M. Berry and Michael Dieter (eds.) *Postdigital Aesthetics. Theory, Culture and Society.* Available at: www.theoryculturesociety.org/review-of-david-m-berry-and-m ichael-dieter-eds-postdigital-aesthetics-by-m-beatrice-fazi/ [Accessed 18 January 2020].

Feenberg, A., 2019. Postdigital or Predigital? *Postdigital Science and Education*, 1 (1), pp.8–9.

Hammond, L., 2021. Children, childhood, children's geographies: Evolving through technology. In N. Walshe and G. Healy (eds), *Geography Education in the Digital World.* London: Routledge, pp.38–49.

International Geographic Union, Commission on Geographical Education (IGU-CGE), 2016. *International Charter on Geographical Education.* International Geographical Union, Commission on Geographical Education. Available at: www.igu-cge.org/wp-content/uploads/2019/03/IGU_2016_eng_ver25Feb2019.pdf [Accessed 31 May 2019].

Kerski, J.J., 2015. Opportunities and challenges in using geospatial technologies for education. In O. Muniz Solari, A. Demirci and J. van der Schee (eds), *Geospatial Technologies and Geography Education in a Changing World: Advances in Geographical and Environmental Sciences.* Tokyo: Springer, pp.183–204.

Kitchen, B., 2021. Using mobile virtual reality to enhance fieldwork experiences in school geography. In N. Walshe and G. Healy (eds), *Geography Education in the Digital World.* London: Routledge, pp.131–141.

Lambert, D., 2015. Research in geography education. In G. Butt (ed.), *Masterclass in Geography Education: Transforming Teaching and Learning.* London: Bloomsbury, pp.15–30.

Lambert, D., 2018. Teaching as a research-engaged profession: Uncovering a blind spot and revealing new possibilities. *London Review of Education*, 16 (3), pp.357–370.

Lambert, D. and Biddulph, M., 2015. The dialogic space offered by curriculum-making in the process of learning to teach, and the creation of a progressive knowledge-led curriculum. *Asia-Pacific Journal of Teacher Education*, 43 (3), pp.210–224.

Lambert, D. and Solem, M., 2017. Rediscovering the teaching of geography with the focus on quality. *Geographical Education*, 30, pp.8–15.

Lambert, D., and Walshe, N., 2018. How Geography Curricula Tackle Global Issues. In A. Demirci, R. Miguel González and S. Bednarz (eds), *Geography Education for Global Understanding.* London: Springer, pp.83–96.

Massey, D., 2005. *For Space.* London: Sage.

Morgan, J., 2019. *Culture and the Political Economy of Schooling: What's Left for Education?* Abingdon: Routledge.

Muniz Solari, O., Demirci, A. and van der Schee, J.A., 2015. *Geospatial Technologies and Geography Education in a Changing World: Geospatial Practices and Lessons Learned – Advances in Geographical and Environmental Sciences.* Tokyo: Springer.

Pike, S., 2021. GIS for young people's participatory geography. In N. Walshe and G. Healy (eds), *Geography Education in the Digital World.* London: Routledge, pp.117–128.

Priestnall, G., 2021. Augmented reality: Opportunities and challenges. In N. Walshe and G. Healy (eds), *Geography Education in the Digital World.* London: Routledge, pp.155–167.

Puttick, S., 2015. Chief examiners as prophet and priest: Relations between examination boards and school subjects, and possible implications for knowledge. *The Curriculum Journal*, 26 (3), pp.468–487.

Puttick, S., 2021. Digital technologies and their roles in knowledge recontextualisation and curriculum making. In N. Walshe and G. Healy (eds), *Geography Education in the Digital World.* London: Routledge, pp.17–25.

Roberts, M., 2021. Geographical sources in the digital world: Disinformation, representation and reliability. In N. Walshe and G. Healy (eds), *Geography Education in the Digital World.* London: Routledge, pp.53–64.

Schaal, S., 2021. Location-based games for geography and environmental education. In N. Walshe and G. Healy (eds), *Geography Education in the Digital World.* London: Routledge, pp.168–178.

Taylor, L., 2013. The case as space: Implications of relational thinking for methodology and method. *Qualitative Inquiry*, 19 (10), pp.807–817.

Tesar, M. and Hood, N., 2019. Policy in the time of the Anthropocene: Children, childhoods and digital worlds. *Policy Futures in Education*, 17 (2), pp.102–104.

Walshe, N., 2017. Developing trainee teacher practice with geographical information systems (GIS). *Journal of Geography in Higher Education*, 41 (4), pp.608–628.

Walshe, N., Driver, P. and Keenoy, M., 2021. Navigating the theory-practice divide: Developing trainee teacher pedagogical content knowledge through 360-degree immersive experiences. In N. Walshe and G. Healy (eds), *Geography Education in the Digital World.* London: Routledge, pp.26–37.

Index

Page numbers in italics refer to figures. Page numbers in bold refer to tables.

2D photographs 135

academic discipline 18, 38–39, 46
accessibility 20, 42, *43*, 68, 173
accountability 7, 33, 38, 46
Allaway, Richard 54
anthropocene 183
Apple ARKit 157
application programming interfaces (APIs) 79
appropriation of space 42, *43*
approximations of practice 27
AR *see* augmented reality (AR)
ArcGIS Online 93, 96, 102, 103, 104, 106, 107, 112, 120, 121, 123
Artcodes 158
artefacts 15, 18, 53
articles, and newspapers 56
Ash, J. 18
Association for Science Education (ASE) 133
atlases 53
audio-recorded interviews 30
augmented reality (AR) 144, 155, 171; augmented landscape models 162; categories of **156**; codes 157; field-based augmentation 160–161; immersive 156–157; landscape models 160; location-based 158–159; marker-based 157–158; non-immersive 157; objects 159–160; place-specific 158–159; recognisable patterns 158; remote 159; sandboxes 160; sensor-based 155–156; spatially based 159–160; techniques 155–156; *see also* virtual reality (VR)
Avouris, N.M. 172

Baker, T.R. 90, 93
Ball, S. 8
banking education 39
Barnett, J. 27
Barnett, R. 10

Bartie, P.J. 159
Bartlett, J. 53
Baston, J. 97
BECTA report 55, 57
Bednarz, S. 117
'being a professional' 10
Béneker, T. 112
Berglund, U. 120
Berners-Lee, Tim 53, 55
Biddulph, M. 125
binary computing architectures 18
BioDiversity2GoSimulation game 173
Bisphan, J. 13
Bouwmans, M. 112
Bovill, M. 40
British Educational Research Association (BERA) Ethical Guidelines 30
Brooks, Clare 2, 7, 26, 32, 90, 92, 98
Brown, G. 44
Browne, K. 44
Butt, G. 182, 183

Cadwallader, T. 95
Cairncross, F. 78
Cambridge Assessment 57
capitalism 42
Carbonell, C. 157
Carswell, J.D. 159
Carter Review 26, 27
CERN 53
Chang, Chew-Hung 3, 142, 147
Chatterjea, K. 149
children: and childhood 39–40; engaging with digital world 41
children's geographies 38–39; defined 39–40; digital world, children's narrative 43–45; in geography education 45–46; geography education in schools 41–43; social identities 39; social spaces 43–44; technology in 40–41

citizen science 76
Clandinin, D.J. 9
Clarke, E., 65
Clarke, T. 65
classroom surveillance 7
Clifford, N. 172
climate-smart decisions 57
cloud generators 82
codes 157
cognitive activation 171
cognitive thinking 29, 148
communications technologies 78
computer-generated content 145
computers in school 53–55
Connecting the Classrooms (CTC) 65–66, **69**; digital pedagogies in higher education 66–67; in-class reflection 70; introduction 68; learning outcomes *70*; online interaction 69–70; pre-CTC set up 68; reflection 70–73; research time and presentations 68–69; stages of 68–70; tasks and activities 67–70
Connelly, F.M. 8, 9
contact with strangers 40
content analysis techniques 82
continued professional development (CPD) 90
contributed geographical information (CGI) 118
Cook, V. 134
Counsell, C. 90
Crawford, K. 78
Cristianini, N. 77
critical conscience 42
critical thinking 67
Crutcher, M. 79
CTC *see* Connecting the Classrooms (CTC)
curriculum: artefact 104; curation 14–15; geography 10; making 1, 2, 10, 18, 20, 23, 24, 38, 46; social nature of 11; teachers' knowledge about 28; schools 120
customised web mapping applications 104
cyberbullying 40

Danakil Depression 111
Danby, S. 120
data privacy 80–81
decision-making 169
decolonizing knowledge 71
Delors, J. 142, 148
DigComp 2.0 **61**
Digimap 93
digital augmentation 162
digital competence 61
Digital Competency framework for Citizens **61**
digital fluency 61
digital games 66; *see also specific games*
digital geographies 18–19, 21–23
digital information 24

digital literacy 61
digital pedagogies: challenges 65; in higher education 66–67
digital technologies 17–18; curriculum making 21; in geography 18–19; information accessed through media 21–23; intensification of teachers' work 21; space-times of teachers' journeys for information 19–21
disaster and crisis management 77
distanciation 42, *43*
domination of space 42, *43*
Drexel University 67
Driver, Paul 2, 26, 29
drone-based learning 66

Earle, P.S. 77
earthquakes 55; detection method 77; magnitudes and depth in ArcGIS Online *106*; patterns 102
EcoMOBILE project 159
EDINA Mapstream 110
editable feature services (EFS) 103
educational geogames 169–170, 172
EduTwitter 13
Efstathiou, I. 158
Egiebor, E.E. 120
Ellis, V. 98
Elwood, S. 125
environmental quality fieldwork data *108;* Esri Survey 123 **109**
Es, E.A. van 27
European Environment Agency 56, 60

Facebook 75, 76, 118, 119
Facebook/Cambridge Analytica scandal 81
'face-to-face' remote engagement 72
fake news 60–61
Fargher, Mary 3, 91,102, 125, 147
Favier, T. 149
Fearnley, Francesca 3, 75, 107
Ferguson, Rebecca 66
field-based augmentation 160–161
Field Studies Council (FSC) 133
fieldwork, in geography 131–132; literature on 147; phases of 133
film-strips 53
Firth, R. 113
flat video 28
Flickr 75
'flipped classrooms' 65
floods and cyclones 55
forbidden spaces 44
formal educational systems 12
Foster, E.J. 120
fracking in QGIS Cloud **110**
Freire, P. 39

friction of distance 44
Friedman, Thomas 78
Frøyland, M., 133

GA *see* Geographical Association (GA)
game-related approaches 168
GapMinder 55, 57
GE *see* Google Expeditions (GE)
General Data Protection Regulations (GDPR) 81
generative growth 27
GeoCapabilities 46, 104
geogames 168; design *172*, 172–173; educational 169–170; and educational outcomes 170–172
Geographical Association (GA) 54, 89, 133; Annual Conference 97; fieldwork 131
geographical decision-making 118
geographical enquiry 19, 58, 59, 94, 97, 102, 112, 117, 120-122, 131, 135-137, 148, 149
geographical information systems (GIS) 3, 89, 102, 144, 149, 160; academic literature 94–96; characteristics of professional discourse 92–94; geography curricula in England 89; research approach 92; in school 118; teachers' preparedness 91–92; in teachers' professional development 90; teachers' professional practice 96–97
geographical sources 53; computers in schools 53–56; in digital world 56–57; opportunities for teachers and learners 57–59; teachers and learners, challenges 59–62; World Wide Web 54–56
geographies of development 65
geographyalltheway 55
geography education 54, 75; community 10, 89; in digital world 181–182; in post-digital world 183; research 19
geography fieldtrips 148
Geography National Curriculum 10, 20, 96
geographypods 54
geography teachers and GIS 2, 7, 102–103; earthquake patterns in Esri ArcGIS Online 105–107; environmental quality data with Esri Survey 123 107–110; Esri StoryMap to develop knowledge of Danakil, Ethiopia 111–112; powerful disciplinary knowledge (PDK) 104–112; QGIS Cloud, fracking in 110–111; WebGIS 102–104
geo-located data 103
geospatial technologies 109, 120, 182
geotag 78
Giddens, A. 119
Gill, L. 159
GIS *see* geographical information systems (GIS)

GIS for young people's participatory geography 117–118; Connecting to local 120–121; decision-making 120–121; geographical learning 121–122; geospatial technologies 118; and local places 118–120; potential of 124–125; Sacred Heart Boys' National School 122–123; St Kevin's College 123–124; *see also* geographical information systems (GIS)
Glesinger, K. 97
'Global Classroom' 67
global positioning system (GPS) 158
Goodchild, M.F. 76, 117
Goodson, I. 8, 41
Google Classroom 181
Google Earth 55, 56, 120
Google Expeditions (GE) 3, 132; attitudes and values 137; awareness 136; different locations 137–138; enquiry questions for investigation 135–136; evaluation 137; familiarisation with locations of fieldwork 134; fieldwork in geography 133–134; knowledge and understanding 136–137; locations and habitats 134; observation skills 135; physical fieldwork 132–133, 136; post-physical fieldwork 137; pre-physical fieldwork 133; risk assessment 134–135
Google Maps 56, 110
Google ARCore 157
Google search 22–23
GPS positioning 156
Greater London Authority 56
Grossman, P. 27

Hague, C. 57
Haigh, Martin 67
Haigh's indicators of IoC 71–72
Hammond, Lauren 2, 33, 38, 119, 183
Hart, R. 119, 124
Harvey, D. 42
Harvey's grid of spatial practices 42, *43*
Hawley, D. 106
Haynes, S. 135
head mounted displays (HMD) 156
Health Protection Agency 77
Healy, Grace 1, 3, 89, 96, 102, 103, 125, 147, 181
Heath, R. 95, 103
Heidegger's notion of being 10
Hennig, B. 96
Heritage Foundation 60
Hesslewood, A. 96
Hew, K.F. 76
higher education (HE), teaching of geography in 3, 65; *see also* geography education
Hodson, D. 27
Höhnle, S. 96

HoloLens 156
Hood, N. 183
HS2 136–137
Huizinga, J. 170
human geographies 20
Humans of New York (HONY) project 66–67
Humphreys, L. 157
hurricane *59*
Hwang, G.J. 155, 157
hyperlinked multimedia information 65
hyper-socialisation 7, 10, 12

ideal identity 13
immersive AR 156–157
immersive overlay platform (IOP) 30
in-field VR 161
information about crimes 44
Ingress 168
initial teacher education (ITE) 26, 90
innovative pedagogy 66
Instagram 75, 76
Institute for Economic Affairs (IEA) 60
Institute for Public Policy Research (IPPR) 60
Intergovernmental Panel on Climate Change (IPCC) 56
inter-group empathy 71
international education 67
International Geographical Union – Commission on Geographical Education (IGU-CGE) 142, 182
internationalisation of curriculum (IoC) 66, 67, 71–72
Internet 11, 14, 41, 53-56, 75; *see also Web 2.0*
intuitive mapping 103
IT management system 75
Iwaskow, I. 96

Jacobs, V. 27
Jamali, S.S. 157
Jarvis, C. 149
Jo, I. 90
'just-in-time' teaching 65

Kamarainen, A.M. 171
Keenoy, Mandy-Jane 2, 26
Kishino, F.A. 155
Kitano, M. 67
Kitchen, Rebecca 3, 131
Klopfer, E. 159, 171

Lambert, D. 33, 38, 46, 113, 131
Lampos, V. 77
landscape models 160
Lange, E. 159
Layar 157
LBG *see* location-based games (LBG)

learning objectives (LOs) 10–11
Leask, B. 67
Lefebvre, Henri 42
Liao, T. 157
LinkedIn 75
Living Atlas 104
Livingstone, S. 40
location-based AR 158–159
location-based games (LBG) 168; educational geogames 169–170; geogame design 172–173; geogames and educational outcomes 170–172; trends 173–174
location-based multimedia 158
location-based technology 119
location-triggered multimedia 155
locative media *161*
Lynch, K. 119

Mackaness, W.A. 159
marker-based AR 157–158
Maskall, J. 133
Massey, D. 119
mass media 20
material technologies 18
Maude, A. 104, **107,110,111**
Mayer, R.E. 162
McDonald, M. 27
McDougall, D. 134
McEwan, L. 65
McMahon, G 119
meaning-making 39
Meek, S. 159
mental maps, elements of 121–122
Milgram, P. 155
Miller, C. 53
Minocha, S. 135
Mitchell, David 7, 10, 11, 55, 103
mixed-reality 169
Mobile AR 157
mobile complex 11
mobile devices 12
mobile electronic devices (MEDs) 170
mobile geospatial technologies 103
mobile learning model 14
mobile location-based learning 168
mobile mass communication 12
mobile technologies and fieldwork 142–143, 149; fieldwork and geographical learning 147–148; in geography education 143–145; technological pedagogical content knowledge 145–147; trends 143–144, **143**, *144*
mobile virtual reality: fieldwork in geography 131–132; Google expeditions 132–138; issues and implications 138; virtual reality 132; *see also* virtual reality (VR)
monoscopic camera 29

Moreno, R. 162
Morgan, J. 38, 46
Morrill, R. 18
multi-academy trusts (MATs) 33, 57
multi-cache 169
multi-tasking 65
Muniz Solari, O. 182

National Curriculum in England 20
National Digital Strategy 121
naturalistic coding 30
near-field communication (NFC) 169
neogeographers 118
networked communities 9
non-governmental organisations (NGOs) 39–40
non-immersive AR 157
Nordin, K. 120

Oers, B. van 31
Ofsted 33, 34n1; Education Inspection Framework 33
omnidirectional recording 29
online education platforms 181
online maps 55
online photographs 56
online resources 60
online social media engagement 14
online social spaces 2, 13–14
online social structure 14
open source data 103–104
Open StreetMap 110
Ordnance Survey maps 53, 110
overhead transparencies 53

Pachler, N. *12*
Padfield, Rory 2–3, 65
parallel processing 65
Parkinson, A. 138
participation dimension **58**
participatory geographical learning 117–118
Payton, S. 57
pedagogical content knowledge (PCK) 26, 27–29, 145
pedagogical decision-making 32
performativity 38
Perry, J. 171
Persaud, I. 89
pervasive games 169
photo-packs 53
physical fieldwork experience 133
Pike, Susan 117
Pinterest 75
place-specific AR 158–159
Podbury, Matt 54
PokémonGo 168
policy makers 39

Pollard, G. 96
pornography 40, 44, 45
port of call 20
post-fieldwork reflective phase 133
post-physical fieldwork phase 137
post-teaching teacher reflection 30
powerful disciplinary knowledge (PDK) 104–112, **105**; earthquake patterns analysis in Esri ArcGIS Online 105–107, **107**; environmental quality data and Esri Survey 123 107–110; Esri StoryMap to develop knowledge of Danakil, Ethiopia 111–112; fracking in QGIS Cloud 110–111
pre-fieldwork induction 133
Priestnall, Gary 3, 155, 182, 183
Primary Geography (PG) 92, **93**
Primary School Curriculum 121
problem-solving 169
production of space 39
Production of Space, The (Lefebvre) 42
professional identity 9, 13
professional knowledge landscapes 9
professional learning 12, 90
professional normalisation 13
professional vision 27
Projection Augmented Relief Model (PARM) 160, 162, *163*
Puttick, Steve 2, 10, 17, 55

QGIS Cloud 103, 106
quick response (QR) codes 157–158

Rawding, C. 131
recognisable patterns 158
recontextualisation 1, 2, 18–19, 24
Redecker, C. 57
Refugee Council 55
Reiss, M. 131
Remmen, K. 133
remote 159
representational space 42
Ricker, B. 93
Roberts, Margaret 2, 21, 53, 145, 148, 149, 182
Robson, J. 13, 14
rock specimens 53
Royal Geographical Society with Institute of British Geographers (RGS-IBG) 54, 103
Ruchter, M. 170

Sachs, J. 9
Sacred Heart Boys' National School 122–123
Schaal, Steffen 3, 168, 171, 182
school-centred training, in teacher preparation or ITE programmes 26–27, 28
schools: geography 39; geography education in 41–43; policies 7

search engine algorithms 20
search engine marketing (SEM) 61
search engine optimisation (SEO) 61
self-efficacy 32
sensor-based AR 155–156, 159
SentiStrength 82
sequencing of location-based tasks *173*
sexually transmitted infections (STIs) 45
sexual messaging ("sexting") 40
Sheldon, J. 171
Shelton, T. 78
Sherin, M.G. 27
Shin, E.E. 117
SimCity 170
Skype 72, 75
slides 53
smart phones 76
Smith, P.K. 40
Snapchat 75, 118, 119
social change 41
social construction, of children 40
social identities 39
social media 12, 14, 40, 41, 75, 118; citizens as sensors 76; data privacy 80–81; GDPR 81; geotag, beyond the 78; limitations of research 78–81; representation 79; sampling 79–80; situational awareness 77; SMGI 81–83; spatial extent of phenomena 77; Twitter and research 76–78
social media geographic information (SMGI) 76, 79, 81–83
social networking 75, 81
social online space 14
social structure 13
socio-cultural ecological approach 11; agency 11, *12*; cultural practices 12, *12*; to mobile learning *12*; structures *12*, 12–13
socio-techno-cultural productions 18
software package 53
Solem, M. 113
spatially based AR 159–160
spatial practices 42
spatial thinking 125
spatial variation 144
springboards 28
Squire, K. 159
Stake, R.E. 29
standard video 28
Star, J.R. 27
stereoscopic video 29
St Kevin's College 123–124
Stokes, A. 133
StoryMaps 93, 95, 102, 104, 111, **111**, 112, 120
storytelling 41–42, 44, 45

sub-cultures 8
subject and pedagogy 26
subject identity 9
subject-knowledge development 26
subject-specificity in lesson observations 29
subject-specific pedagogy 26, 27
Sui, D. 18
Swords, J. 125

Tang, Y. 76
Tani, S. 38
Taylor, L 97, 181
teachers: conservative 13; curriculum curation 14–15; as curriculum makers 10–11, 14; curriculum making 3, 38; education 8; engagement, with digital technologies 2; engagement with digital technologies 21; knowledge recontextualisation 18–19; mobile learning, socio-cultural ecological approach 11–13; professionalism 1; professional knowledge landscape 8–9; professional practice 9–10; and reforms in education 9; subject identity 7; subject specialism 8; vocationally committed 8; *see also* trainee teacher
teacher identity 7; importance of 7–8; online social spaces and 13–14
teacher-to-teacher interaction 2
teaching assistant (TA) 32
Teaching Geography (TG) 54, 92, **93**, 94, 98
technological developments 76
technological pedagogical content knowledge (TPACK) 49, 91, 145–147, *146*, 149
technology enhanced assessment 66
teenage pregnancies 45
Tesar, M. 183
textbooks 53
Thatcher, J. 93
thematic analysis 92
think-aloud protocol 29
3D block diagrams 157
360-degree video 2, 26–27; individual student interviews 30; noticing for understanding 31–33; pedagogical content knowledge 27–28, 32–33; post-teaching teacher reflection 30; project design 29–30; teaching recorded with 29; trainee teachers' lesson observation 31–32; video and immersion 28–29; in virtual reality 30
TicTacToe 173
Tidy City 173
Tilling, S. 131
time-critical situations 77
time-series analysis 82
Tobler's First Law of Geography 78
Trafford, R. 95

trainee teacher: ability to notice 27; decision-making 27; flexible interpretation 27; inexperience of 27; subject knowledge 26
transformative learning 71
transformative curriculum 67
Tudor, A. 136
Turner, A. 118
Twitter 13–14, 15, 75, 76, 77, 79–83

UN Convention on the Rights of the Child (UNCRC) 120
UNESCO 14
United Nations 55
United Nations Educational, Scientific and Cultural Organisation (UNESCO) 142
United States Geological Survey (USGS) Earthquakes Hazards Program 106
University of Leeds 67
'user-generated' content 75
US Geological Survey (USGS) 77; Earthquakes Hazard programme 55

Van Der Schee, J. 149
van Manen, M. 92
video-cassettes 53
video conferencing tools 75
video technologies 26
video viewing 28, 30
virtual environments 28
virtual reality (VR) 144, 155; and 360-degree video 30; Google Expeditions (GE) 131–139; see also augmented reality (AR)
virtual spaces 9
virtual studios 66
VITAE project 8
Vogel, S. 171
volcanic eruptions 55

volunteered geographical information (VGI) 76, 104, 118, 145
VR see virtual reality (VR)
VR headsets 30

wall maps 53
Walshe, Nicola 1, 2, 26, 29, 90, 95, 98, 110, 112, 125, 147, 181
Web 2.0 2, 38–39, 40, 41, 75
web-based investigation 59
Web-based mapping platforms 103
web-based social networking 76
WebGIS 3, 93, 102–104, 108; customised web mapping applications 104; intuitive mapping 103; mobile geospatial technologies 103; open source data 103–104
web map services (WMS) 104
WhatsApp 72
Wikitude 157
Willy, T. 97
Winter, C. 131
World Bank 55
World Health Organisation 55
World Is Flat, The (Friedman) 78
WorldMapper 57
World Wide Web (WWW) 40, 54–56, 75
Wu, B.S. 145, 147
Wu, P.H. 155

Yiannoutsou, N. 172
Young Scientist Competition 120
YouTube 44, 56

Zimmer, M. 79, 80
Zombies 168, 169
Zook, M. 79
Zoom 72
Zuckerberg, M. 81

For Product Safety Concerns and Information please contact our EU
representative GPSR@taylorandfrancis.com
Taylor & Francis Verlag GmbH, Kaufingerstraße 24, 80331 München, Germany

www.ingramcontent.com/pod-product-compliance
Lightning Source LLC
Chambersburg PA
CBHW060343010526
44117CB00017B/2949